THE EXPERIENCE OF TIME

CONTROVERSIES IN PSYCHOANALYSIS SERIES

Identity, Gender, and Sexuality: 150 Year after Freud
edited by Peter Fonagy, Rainer Krause, & Marianne Leuzinger-Bohleber

Transvestism, Transsexualism in the Psychoanalytic Dimension
edited by Giovanna Ambrosio

THE EXPERIENCE OF TIME

Psychoanalytic Perspectives

edited by

Leticia Glocer Fiorini
and Jorge Canestri

Foreword by Henry F. Smith

Controversies in Psychoanalysis Series

KARNAC

First published in 2009 by
Karnac Books Ltd
118 Finchley Road, London NW3 5HT

British Library Cataloguing in Publication Data

A C.I.P. for this book is available from the British Library

ISBN 978 1 85575 775 2

Edited, designed and produced by The Studio Publishing Services Ltd,
www.publishingservicesuk.co.uk
e-mail: studio@publishingservicesuk.co.uk

Printed in Great Britain

www.karnacbooks.com

CONTENTS

Controversies in Psychoanalysis Series

IPA Publications Committee

The present Publications Committee of the International Psychoanalytical Association continues with this title to expand the Controversies in Psychoanalysis series, the objective of which is to reflect, within the frame of our publishing policy, present debates and polemics in the psychoanalytic field.

Theoretical and clinical progress in psychoanalysis continues to develop new concepts and to reconsider old ones, often in contradiction with each other. By confronting and opening these debates, we might find points of convergence, but also divergences that cannot be reconciled; the ensuing tension among these should be sustained in a pluralistic dialogue.

The aim of this series is to enlighten these complex intersections through various thematic proposals developed by authors from within different theoretical frameworks and from diverse geographical areas, in order to open possibilities of generating a productive debate within the psychoanalytic world and related professional circles.

The present title focuses on the experience of time, a subject that generates deep controversies from different points of view: psychoanalytic, cultural, social, and ethical. The contributors have accepted the challenge to consider and display the plural, heterogeneous dimensions of time that involve the analytic relation as well as the development and construction of subjectivity.

We are pleased to continue this series with the support of Cláudio Eizirik, President of the International Psychoanalytical Association. Special thanks are due to the contributors to this volume.

Leticia Glocer Fiorini
Chair of the Publications Committee

ABOUT THE EDITORS AND CONTRIBUTORS

Ingeborg Magda Bornholdt, MSc, psychologist, is a member and training analyst of Porto Alegre Psychoanalytical Society, and a child and adolescent psychoanalyst. Her paper "Construções da temporalidade no desenvolvimento normal" (Constructions of temporality in normal development) received the Fepal (Latin America Psychoanalytical Federation) prize for child and adolescent analysis in 2002.

Jorge Canestri, MD, is a psychiatrist and psychoanalyst, training and supervising analyst for the Italian Psychoanalytical Association (AIPsi) and for the Argentine Psychoanalytic Association (APA). He is a Full Member of the International Psychoanalytic Association, and the Mary S. Sigourney Award recipient, 2004. He is Chair of the Working Party on Theoretical Issues of the EPF, Member of the Conceptual and Empirical Research Committee (IPA), and Professor of Psychology of Health at the Roma 3 University. In addition, he is Editor of the Educational Section of the *International Journal of Psychoanalysis*, a member of the Editorial Board of the *International Journal of Psychoanalysis*, IPA Global Representative for Europe 2005–2007, Representative for Europe to the Executive Committee

2007–2009, and President of the Italian Psychoanalytical Association. He has published numerous psychoanalytical papers in books and reviews, and is co-author of *The Babel of the Unconscious. Mother Tongue and Foreign Languages in the Psychoanalytic Dimension*. He edited (with Marianne Leuzinger-Bohleber and Anna Ursula Dreher) *Pluralism and Unity? Methods of Research in Psychoanalysis*, and is editor of *Psychoanalysis: From Practice to Theory* (New York: Wiley). He also edited, with G. Ambrosio and S. Argentieri, *Language, Symbolization, and Psychosis* (London: Karnac). He is Director of the webpage: "Psychoanalysis and logical mathematical thought".

Leticia Glocer Fiorini is a training psychoanalyst of the Argentine Psychoanalytic Association. She is the current chair of the Publications Committee of the International Psychoanalytical Association, current chair of the Publications Committee of the Argentine Psychoanalytic Association, and former member of the Editorial Board of the *Revista de Psicoanalisis* (1998–2002, Buenos Aires). She won the Celes Cárcamo Prize (APA, 1993) for her paper: "The feminine position: a heterogeneous construction". She is the author of *Deconstructing the Feminine. Psychoanalysis, Gender and Theories of Complexity* (London: Karnac), and co-editor of *On Freud's "Mourning and Melancholia"* (London: Karnac). She edited, in Spanish, *El Otro en la Trama Intersubjetiva* (The Other in the Intersubjective Field), *Tiempo, Historia y Estructura* (Time, History and Structure) and *Los Laberintos de la Violencia* (Labyrinths of Violence). She has also published numerous papers and chapters in books about femininity.

André Green is a psychoanalyst who lives and works in Paris. He is Past-President of the Paris Psycho-Analytical Society. He also served as Director of the Paris Psychoanalytic Institute and as Vice-President of the International Psychoanalytic Association. In addition, he has been co-editor of the *International Journal of Psychoanalysis, The International Review of Psychoanalysis*, and *Nouvelle Revue de Psychanalyse*. He has written several books, which include *On Private Madness, The Work of the Negative; The Fabric of Affect in the Psychoanalytical Discourse; The Chains of Eros; Life Narcissism. Death Narcissism*; and *Key Ideas for a Contemporary Psychoanalysis*.

Charles Hanly is training and supervising analyst in private practice, a member of the Canadian Psychoanalytic Association (Toronto Branch) and an Emeritus Professor of Philosophy at the University of Toronto. He is the author of numerous articles and books on psychoanalytic theory, clinical and applied psychoanalysis. He is President-elect of the International Psychoanalytic Association.

Otto F. Kernberg, MD, FAPA, is Director of the Personality Disorders Institute at The New York Presbyterian Hospital, Westchester Division, and Professor of Psychiatry at the Weill Medical College of Cornell University. Dr Kernberg is a Past-President of the International Psychoanalytic Association. He is also Training and Supervising Analyst of the Columbia University Center for Psychoanalytic Training and Research. He was awarded the 1972 Heinz Hartmann Award of the New York Psychoanalytic Institute and Society, the 1975 Edward A. Strecker Award from the Institute of Pennsylvania Hospital, the 1981 George E. Daniels Merit Award of the Association for Psychoanalytic Medicine, the 1982 William F. Schonfeld Memorial Award of the American Society for Adolescent Psychiatry, the 1986 Van Gieson Award from the New York State Psychiatric Institute, the 1987 and 1996 Teacher of the Year Award from The New York Hospital-Cornell Medical Center, Westchester Division, and the 1990 Mary S. Sigourney Award for Psychoanalysis. He was elected to membership of the Society of Scholars of the Johns Hopkins University in 1992. He received the 1993 I. Arthur Marshall Distinguished Alumnus Award, Menninger Alumni Association, The Menninger Foundation, the 1993 Presidential Award for Leadership in Psychiatry from the National Association of Psychiatric Health Systems, and the Distinguished Service Award from the American Psychiatric Association in 1995. He was elected Doctor Honoris Causa by the University of Buenos Aires, Argentina, in 1998, and received the Austrian Cross of Honor for Science and Art in 1999. In 2007, he received the Golden Medal of Honor to the City of Vienna, and the Thomas William Salmon Medal from the New York Academy of Medicine, New York. He is the author of nine books and co-author of fifteen others.

José E. Milmaniene is a psychiatrist and psychoanalyst. He is a member of the Psychoanalytical Association of Argentina and its present scientific secretary. Among his books are, *The Ethic of the Subject; The Place of the Subject; The Subject's Time; The Function of the Father; The Holocaust; Strange Couples; Clinical Text. Kafka, Benjamin, Levinas.*

Michael Parsons is a training and supervising analyst of the British Psychoanalytical Society. He studied medicine and specialized in psychiatry after a first degree in Classics and Philosophy. He works in private psychoanalytic practice in London and is well known in the UK and internationally as a lecturer and seminar leader. He is the author of *The Dove that Returns, The Dove that Vanishes: Paradox and Creativity in Psychoanalysis* (Routledge, 2000) and co-editor of the collected papers of Enid Balint under the title *Before I was I: Psychoanalysis and the Imagination* (Free Association, 1993).

Rosine Jozef Perelberg is a training analyst and supervisor, Fellow of the British Psychoanalytical Society, and Visiting Professor in the Psychoanalysis Unit, Division of Psychology and Language Sciences at University College London. She was Chair of the Curriculum Committee of the British Society and served on the Admissions and Education Committees. In 1991 she was co-winner of the Cesare Sacerdoti Prize at the International Psychoanalytical Association Congress in Buenos Aires. She co-edited, with Joan Raphael-Leff, *Female Experience: Four Generations of British Women Psychoanalysts on Work with Women* (1997, second edition 2008). She has edited *Psychoanalytic Understanding of Violence and Suicide* (1998); *Dreaming and Thinking* (2000, 2003); *Freud: A Modern Reader* (2005); and *Time and Memory* (2007). She has written *Time, Space and Phantasy* (2008). She works in London, in private practice.

Janine Puget, MD, psychoanalyst, is a Full Member and training analyst of the Asociación Psicoanalítica de Buenos Aires (APdeBA), IPA, Federación Psicoanalítica de América Latina (FEPAL), Founding Member and Honorary Member of the Asociación Argentina de Psicología y Psicoterapia de Grupo (Argentine Group Psychology and Psychotherapy Association) (AAPPdeG), Member of the Executive Council, Asamblea Permanente por los Derechos Humanos (Permanent Human Rights Assembly) (APDH), and Co-Director of Magister Family and Couple Psychoanalysis IUSAM-APdeBA. Her publications include: *Lo vincular. Teoría y Clínica psicoanalítica* (co-edited with I. Berenstein (Paidós, 1997); *Psychanalyse du lien. Clinique et théorie,* (París, Erès, 2008); *Violence d'etat et psychanalyse* (co-authored with R. Kaës and others) (Dunod, 1989); *Psychoanalysis Amid the Unthinkable: Essays on the Argentinian Experience.* (Free Association, 1990). *Violencia de Estado y Psicoanálisis* (Centro Editor, Buenos Aires, 1991); Editorial Lumen, March 2006; *Violenza di stato e psicoanalisi* (editor, Gnocchi, 1994); *Psicoanálisis de la Pareja Matrimonial* (co-authored with I. Berenstein) (Paidós, Buenos Aires, 1988; *Psicanálise do Casal* (Porto Alegre, Artes Médicas, Brazil, 1993); *El Grupo y sus configuraciones: Terapia Psicoanalítica* (co-authored with M. Bernard, G. Games Chaves, & E. Romano) (Lugar Editorial, Buenos Aires,1982); *Il gruppo e le sue configurazioni. Terapia psicoanalítica* (editor) (Borla, Italy, Octubre, 1996).

Satish Reddy, is Assistant Professor of Clinical Psychiatry & Medicine at The Weill Medical College of Cornell University and at the faculty of Columbia University Center for Psychoanalytic Training and Research. He is attending physician in the Departments of Psychiatry & Medicine at New York Hospital, Queens. He has published the following: (with F. Kasmin & U. Mathur) "Syphilitic gastritis in an HIV-infected individual". *American Journal of Gastroenterology, 87*: 1820–1822; (with D. Shimbo et al.) "Exaggerated serotonin-mediated platelet reactivity as a possible link in depression and acute coronary syndromes", *Journal of the American College of Cardiology, 89*: 331–334; (with N. S. Nobler) "Dangerous hyperglycemia associated with electroconvulsive therapy". *Convulsive Therapy, 12*: 99–103; "Psychoanalytic process in a sacred Hindu text, *The Bhagavad Gita,* in *Freud on the Ganges,* edited by Salman Akhtar (The Other Press, 2005); "Psychoanalytic reflections on the sacred Hindu text, *The Bhagavad-Gita,* in *Does God Help? Clinical & Therapeutic Aspects of Religious Belief* (edited by S. Akhtar & H. Parens (Jason Aronson, 2001); (with S. A. Mosovich & D. Shimbo) "Coronary artery disease and depression". *Primary Psychiatry, 7*: 69–75; "Mother–son incest in the movies". *Mind & Human Interaction, 9*: 72–81; "Physical diagnostic signs in critical care medicine: the neurological examination. In: *Infectious Diseases in Critical Care Medicine* (edited by B. A. Cunha) (Marcel Dekker, 1998); "A first look at the dreams of E. E. Cummings: the preconscious of a synesthetic genius". Reporter, *Bulletin of the Association for Psychoanalytic Medicine, 33*: 73–77.

Jean-Claude Rolland is a Full Member and Ancient President of the French Psychoanalytic Association, Director, with Catherine Chabert, of the review *Libres Cahiers pour la Psychanalyse Paris*, author of *Guérir du mal d'aimer* (Gallimard, connaissance de l'inconscient, 1998), and *Avant d'être celui qui parle* (Gallimard, 2006).

Henry F. Smith, MD, is Editor-in-Chief of *The Psychoanalytic Quarterly*; Training and Supervising Analyst at the Psychoanalytic Institute of New England, East (PINE) in Boston; and Chair of the Program Committee of the American Psychoanalytic Association. The author of over 100 papers on the theory and practice of psychoanalysis, recent articles include "Leaps of faith: is forgiveness a useful concept?" published recently in the *International Journal of Psychoanalysis*; "Analyzing disavowed action: the fundamental resistance of analysis", which he gave as a Plenary Address to the American Psychoanalytic Association and published in *Journal of the American Psychoanalytic Association* in 2006, and "Hearing voices: the fate of the analyst's identifications", which was awarded the 2001 Journal Prize of the *Journal of the American Psychoanalytic Association*.

FOREWORD

"The past is present, isn't it?"

Henry F. Smith

Towards the end of the second act of Eugene O'Neill's *Long Day's Journey into Night,* James Tyrone is alone on stage with his morphine-addicted wife, Mary. In real life, Tyrone was James O'Neill, the playwright's father and a well-known actor. Tyrone implores her, "Mary! For God's sake, forget the past!" And, in a moment of rare lucidity, in which she could be speaking for Eugene O'Neill and his entire tragic vision, Mary responds, "Why? How can I? The past is present, isn't it? It's the future, too" (O'Neill, 1956, p. 87).

The past is present, isn't it? Does that make the past timeless, the present and the future its prisoner, as Mary feels? Or is it the other way around: the present and future imprisoning the past? It all depends on your point of view, and much of psychoanalytic history, including all of the papers in this volume, implicitly or explicitly take a stand on this issue. The debate began with Freud—an argument he characteristically had with himself—which is why most analysts can trace their position to him even if they completely disagree with each other. Think of Freud's insistence on the power of memory and its persistence ("Hysterics suffer mainly from reminiscences" [Freud & Breuer, 1895d, p. 7]) and then think of his discovery only four years later of the screening function of memory

("Memories *relating to* our childhood may be all that we possess" [Freud, 1899a, p. 322, original italics]), the one pointing to a relentless past, the other to a continual remaking in the present, a perpetual here-and-now, we might say. At times, the latter appears to be a hard-wired resistance to memory itself, including the memory of any original experience.

Freud keeps weaving both of these points of view into his thinking: the past with a life of its own inside us, the ego resisting, determined to create its own life, which, willy-nilly, turns out yet again to bear the stamp of the past. As Phillips (2006) puts it, "The ego is utopian, but the past keeps giving birth to itself" (p. 83).

But there is another problem here. Since we know we cannot find "the 'events' of the afternoon on the staircase" (Kris, 1956, p. 73), and *après coup* tells us that all experience is plastic, altered by future imperatives, do we not have to question whether we have any experiences at all to remember? Perhaps they were always only *potential experiences* (Smith, 2008).

Consider how quickly one's perception of an everyday event—a chance encounter on the sidewalk, a dinner conversation—changes as soon as the moment passes, the memory of it shaped in every subsequent moment by the current internal context. Sometimes, the change is dramatic: she loves me; she is interested only in herself. Sometimes, it is trivial: he was casually dressed; he was dressed to impress. And it is always personal: I loved seeing her; I was hurt by her tone of voice. The original experience is almost immediately lost in its retranscription, the past becoming present as soon as it is past.

Several of the authors in this monograph are radically opposed on whether it is best to pursue the past or the present. I will mention just a few. Thus, André Green, in describing the importance of the reconstruction of infantile phantasy, decries the current emphasis on the here-and-now that slights the more important knowledge of the there-and-then. Puget, on the other hand, sees the focus on history as a defensive retreat to the past from what she calls the *evental present,* the latter a kind of radical here-and-now that grants maximum potential for creative transformation.

It would seem that Puget's evental present, like Bion's lack of memory, history, and desire, while admirable, can only be approached asymptotically, for is not all perception, like experience, retrospective as soon as we become conscious of it? The moment we are aware of "being in the present", that present is already past. Our self-conscious lives are a series of still photos, like the frames of a film, on which we impose a sense of continuous flow until the final frame catches us unawares.

In another implicit argument in this volume (and there are many), Green masterfully describes the metapsychological implications of Freud's focus on timelessness. But Hanly, in a *tour de force* of critical reasoning, argues that Freud did not mean what Green thinks he meant at all, and, to the extent that Freud was waxing philosophical, he was wrong, wishfully avoiding the march of death by creating a misguided theory of timelessness. In other words, his theory was an enactment of his fear.

Finally, on the matter of time *vs.* timelessness itself, Parsons is the peacemaker, advocating independently that the analyst stand on a metaphorical bridge between the two.

Notice that each of the authors I have mentioned—and it applies to the others as well—speaks from a different level of abstraction, which means that there is a great deal to learn from the papers in this volume as long as one does not get too caught up in comparing one with another, for then the authors cannot help but talk past each other.

The presence of the past

The presence of the past was brought home to me most forcefully by a patient who kept the past alive in her use of the analytic process even as she appeared to be engaged in the task at hand (Smith, 2006).

An analysand in her early forties, who wishes that I would encourage her sexual excitement rather than focus on its angrier edge, senses one day that I have become momentarily distracted. She asks, "What are you doing?", and then answers her own question. "You are simply adjusting your chair. I am so good. I don't turn around and look." She is commenting—somewhat provocatively, I think—on a defensive "goodness" in her behaviour.

"Looking would be too aggressive," I say.

"It would startle you," she says, and then falls silent. After a minute, she tells me she has just become "aroused".

I say, "Notice that you got aroused just after you felt you had lost me."

"Perhaps," she says, sounding unconvinced. Suddenly, it occurs to me that it was not my distractedness, but my comment that was the precipitant for her arousal, and I say, "Or perhaps your arousal started when I said that looking at me would be too aggressive."

"Yes," she says with rising passion, "The sound of your voice got me excited." And this discovery marks the beginning of a journey in which we explore her wish that I might invite her into forbidden pleasures.

The past is present for sure.

But notice what has happened. The erotic excitement that my patient yearns for me to facilitate is now being played out right before our eyes. She has found a way to experience this excitement, stimulated by the sound of my voice, at the very moment when I am speaking about her aggression, the thing she wishes I would not do. In fact, my effort to identify what excites her only excites her further.

So, here we have a dilemma. If the sound of my voice arouses the erotic experience we are trying to analyse, and I cannot then speak about it without arousing her, what am I to do? My patient cannot think about what I am saying because she is so busy using my words to actualize the wishes we are analysing. Put as Freud did (1914g), she cannot remember the past because she is so completely living it in the present. And this is true whether I pursue her history, focus on the here-and-now, or remain silent. She will find a way to incorporate whatever I do or say into her actualized fantasy, in the present.

Might this not be a reason that work in the here-and-now *must* precede the recovery of the there-and-then? If a patient cannot think—or use the analyst's thinking—because she is so busy doing, must not we examine as carefully as possible how she uses and misuses both her mind and our own, so that we can carve a space for memory, not to mention for thinking itself? Or do you believe Anna Freud (1937) was right when, with the Kleinian cloud on the horizon, she warned that "a technique which concentrated too much on the transference" (p. 27) would overwhelm the ego, which would then be swept into the action. These are the historical extremes that outline the two approaches to time past and time present that,

Green suggests, separate the British school from the French. Where you stand on this point determines whether you feel that to focus on the past would cement my patient's resistance to examining her own role in all this, or whether my here-and-now approach so inflames the transference that she can never reflect on what has brought her to this impasse.

In my view, the distinction between Green and Puget begins to break down at this point. Surely, either the patient or the analyst can use both the past and the present for defensive purposes. The patient's here-and-now use of the analyst is certain to escape scrutiny if the two are bent on reconstruction, but, similarly, no sooner do we think we are in an evental present—to use Puget's term—than we discover the master of disguises, the promiscuous past, to be clothing itself in that very evental present in the service of yet another repetition.

In other words, no matter which approach we favour, the pursuit of the past or the analysis of the present, there will always be an ongoing enactment. In that sense my patient's and my dilemma becomes a prototype for all analytic work. To one extent or another, patients always actualize their wishes at the same time as they agree to analyse them; in fact, actualize them with the only things the analyst has at her disposal, her very words and behaviour. In pursuing their wishes, patients disavow the work, and in doing the work they disavow their wishes. I know no other way to address this double disavowal than to analyse it as it is happening in the real time of the hour (Smith, 2006), even though we can be sure that its analysis is simultaneously being incorporated into the very enactment we are analysing. In fact, analyst and patient are at times so engaged in the dance that we have to ask, with Yeats (1928), "How can we know the dancer [either dancer] from the dance?"

Stopping time

My patient's attempt to weave our work into an actualization that she controls is an effort to eliminate any separateness or distance between us. But it is also an effort to stop time and to transcend mortality, so that there will be no death and no termination. She asserts her immortality by placing us in the aggressive grip of the repetition compulsion.

And this is directly related to the matter of time. For, if mortality is our ultimate enemy, then time is his henchman. "Time wounds all heels," as Groucho Marx (1940) said. Picture the Knight playing chess with Death in *The Seventh Seal* (Bergmann, 1957).

Hanly's insight about Freud and timelessness might suggest that every theory we have about time, both lay and scientific, can be seen as an effort to master time in our endless quest—with my patient—for eternal life.

Wordsworth (1807) may have been modest in his "intimations of immortality," and Shakespeare gently and sadly mocks us when Cleopatra, whose erotic lure had once brought Antony to his knees, cries, "I have immortal longings in me" (Shakespeare, 1608, V, 2: 283–284). But patients know the dilemma of trying to capture time in a bottle.

They do so not merely when they play the game of disavowal, as my patient does above, but more concretely when they toy with the frame: request a change of time, come a bit late, leave a bit early, overstay the end of the hour, or simply procrastinate. He who controls time controls life, or so it is said. And if we "kill time", or at least waste it, maybe it won't catch up with us, as it did with Richard II: "I wasted time and now doth time waste me" (Shakespeare, 1597, V, 5: 49)

Freud (1916a) noted our "demand for immortality" (p. 305) in his most succinct and lovely essay on transience, but he had his own immortal longings that, as Hanly points out, may have shaped his theories. In this respect, the death instinct itself, conceived by Freud in a period of hopelessness, might be seen as a kind of theoretical omnipotence: taking death into his own hands, as if, had we the power of death within us, we would not be subject to it or to our inevitable failures. In Hanly's terms, even Green's brilliant focus on the negative and on absence would be a personal attempt to come to terms with the negative within, the dead mother, the absence of representation.

But here I have a question. In the consulting room, it seems to me very difficult to distinguish the patient who is helpless in the face of an internal absence from one who has some agency over this experience of inner absence, an agency that might be tapped. There are some patients who fill the room with their negativity, as if by doing so they might master, or even destroy, the negative within. Sometimes, this appears to be a defence, a kind of psychic *double negative*, as if by a semantic trick the two negatives might cancel each other out, yielding a positive rather than an ever more implacable negative. It is a trick of mind, a false positive, a disavowal ("It is not an absence within me; it is I who creates my own absence, I who am the hand of death"), and we see it not only in patients, but in analysts who, reluctant to accept the hand of fate, attribute agency to patients where there may be none, often disastrously. This then becomes another effort to defeat death, this time by the wishful attribution of agency.

And so, if some theories attribute agency where there is none, and others keep an Orpheus-like eye (Parsons) on the hand of death (stare death in the face, so to speak), relentlessly focusing on the negative in patients and in ourselves, there is no escape. The fear of death, the effort to be its master, to preserve life and defy mortality, is the most fundamental instinct we possess, and it affects all our thoughts about time and its treatment.

"The past is present, isn't it?" says the playwright. And it always is. I believe that, in analysis, our only hope is to analyse that present and that past at the same time as our very analysing will itself become part of the dance. "How can we know the dancer from the dance?" Our only hope is to analyse the dance, as we are dancing. Or is this simply another wishful attempt to defeat the dance of death?

References

Bergmann, I. (Dir.) (1957). *The Seventh Seal.* Film. Svensk Filmindustri.

Freud, A. (1937). *The Ego and the Mechanisms of Defense* (revised edn). New York: International Universities Press, 1966.

Freud, S. (1899a). Screen memories. *S.E.*, *3*: 301–322) London: Hogarth.

Freud, S. (1914g). Remembering, repeating and working-through. *S.E.*, *12*: 145–156. London: Hogarth.

Freud, S. (1916a). On transience. *S.E.*, *14*: 303–308. London: Hogarth.

Freud, S., & Breuer, J. (1895d). *Studies on Hysteria. S.E.*, 2. London: Hogarth.

Groucho, M. (1940). In: *The Marx Brothers Go West.* Film. MGM.

Kris, E. (1956). The recovery of childhood memories in psychoanalysis. *Psychoanalytic Study of the Child, 11*: 54–88.

O'Neill, E. (1956). *Long Day's Journey into Night*. New Haven, CT: Yale University Press.

Phillips, A. (2006). On not making it up. In: *Side Effects* (pp. 75–100). London: Penguin.

Shakespeare, W. (1597). *The Tragedy of King Richard the Second*. In: G. L. Kittredge (Ed.), *The Complete Works of Shakespeare*. Boston, MA: Ginn, 1936.

Shakespeare, W. (1608). *The Tragedy of Antony and Cleopatra*. In: G. L. Kittredge (Ed.), *The Complete Works of Shakespeare*. Boston, MA: Ginn, 1936.

Smith, H. F. (2006). Analyzing disavowed action: the fundamental resistance of analysis. *Journal of the American Psychoanalytic Association, 54*: 713–737.

Smith, H. F. (2008). Six inventions on unconscious fantasy. *Psychoanalytic Inquiry, 28*: 231–255.

Wordsworth, W. (1807). Ode: intimations of immortality from recollections of early childhood. In: M. H. Abrams (Ed.), *The Norton Anthology of English Literature, Volume 2* (pp. 117–122). New York: W. W. Norton, 1962.

Yeats, W. B. (1928). Among school children. In: *The Collected Poems of W. B. Yeats*. London: Macmillan, 1965.

Introduction

Jorge Canestri and Leticia Glocer Fiorini

It is difficult, if not impossible, to write about time without almost automatically feeling obliged to quote some of the phrases with which great thinkers, in their reflections on this theme, have tried to condense their perplexity. Perhaps the most famous, and, hence, the most frequently quoted, is St Augustine: the philosopher expresses a general conviction that when we refer to time, everyone knows what it is, but we all find it very hard to define.

The concept of time is probably not substantially different from other abstract concepts to which we assign a term: a term that actually includes so many things that none of them in isolation defines it clearly. Perhaps the problem in relation to this particular concept is that "time is of the essence": it is something indispensable that we are compelled to respect.

In the remarkable variety of contents displayed by our use of the concept of time, this certainty is spontaneously accepted and held by all.

However, if our discourse on time is to mean anything, we cannot ignore that, as Stephen Hawking states in the many texts this eminent physicist has devoted to its discussion, our perspectives on time have progressively been modified (Hawking, 1988).

Hawking stresses that a historical review of this concept in physics reveals that, until the early twentieth century, we believed in the existence of absolute time. The word "time" seemed to describe one entity for all events. The discovery of the speed of light, equal for all observers regardless of their movement, leads into the theory of relativity, forcing us to discard the idea of one absolute time. Each observer has an individual measurement of time, measurements that differ from each other. Since the conception of time has become necessarily more complex, our intuitive certainties collide with counter-intuitive theories lacking in uniformity of opinion, even among physicists.

While in "real" time going forward (future) or backwards (past) implies a substantial difference (nobody can remember the future although they may remember the past), we see that the unification of gravity with quantum mechanics recurs to the notion of "imaginary" time, which is confused with direction in space. In imaginary time, unlike real time, forward or backwards makes no difference. For the laws of physics, the distinction between past and future no longer exists when considering the fundamental operations between particles.

However, Hawking always reminds us that in physics we also speak of a "time arrow" in function of the direction indicated by the growth of disorder or entropy. Hawking proposes three different time arrows: the thermodynamic arrow (entropy); the psychological arrow (the brain also obeys the second law of thermodynamics, as Hanly recalls in his paper, mentioning Hawking) and the cosmological arrow (the universe expands and does not contract).

Although there are no divergences in connection with the first "arrow", and, in relation to the second, we have subjective and objective certainty, regarding the third, cosmologists are quite far from agreeing, as demonstrated by the many papers dedicated to this issue, beginning with the debate between Hawking and Penrose.

This debate, included in a six-month programme on the nature of the universe at Cambridge University's Institute of Mathematical Sciences, showcases the differing positions of these two great scientists. Even when their opinions converge, there is palpable divergence in reference to the causes explaining the phenomena described. The discussion of these subjects considerably surpasses the knowledge of most readers of the book produced on the basis of these encounters (Hawking & Penrose, 1996) and exceeds the intentions of our introduction.

It may be useful to recall that the interest of physicists and cosmologists in this controversy has continued with the production of papers and theories and a multiplication of the concepts (and the names) of time, which we need to take into account. "Euclidean" time was added to real time: the former is a time not measured by clocks, but expressed by imaginary numbers that facilitate the calculations that describe what, in quantum mechanics, is called the "quantum tunnel". This modifies the classical conception of time–space in the theory of relativity, since the distinction between time and the three spatial dimensions disappears in favour of four-dimensional space (Vilenkin, 2006).

However, as Bellone (1989) aptly reminds us in his book that examines and reconstructs the history of the concept (in which the temporal order obviously comes into play), natural language "treats the word 'time' as if it were the name of an entity possessing some essential qualities" (p. 10).

He adds that this way of treating the word "time" pertains not only to common sense, but is found in a more sophisticated version in the writings of Galilei, Newton, and even the Einstein of restricted relativity. Space, time, and space–time thus appear as rigid entities with properties independent of the objects of the world. We know today that the perception of time flow, which we all share intersubjectively, cannot stand up to the scientific theories available today regarding time. Einstein himself, with the death of his great friend Michele Besso, writes in a letter to his son and his sister,

> He has preceded me a little in saying goodbye to this strange world. It means nothing. For us who believe in physics, the division between past, present and future has the value of only an obstinate illusion. [quoted by Bellone, 1989, p. 28]

We might question our insistence on the revolution produced by contemporary physics in our conceptions of time. After all, we mentioned the "psychological arrow" and our intersubjective certainty concerning it. If we need to consider time in psychoanalysis, why would this certainty not satisfy us, since our field is psychological rather than physical?

Our hypothesis is that psychoanalysis revolutionizes the common conception of time, similar to the revolution in physics. While we do not ignore the "psychological time arrow", no doubt distinguishing past, present, and future, psychoanalysis reveals that, in analytic experience, time acquires diverse formations in which these distinctions become more complex and fade until they take the shape of what André Green, in a felicitous expression, calls "le temps éclaté" ("exploded time").

In this frame, we understand that analysis of the meanings of the category of time in psychoanalysis involves delimiting the different levels that require study. Every human being is imbued in individual and collective, historical and prospective, repetitive, circular, and open times. Therefore, in this experience, linear, chronological time is only one aspect of the many faces of time.

Psychoanalysis, in theory and practice, cannot be exempt from this imbuement. Precisely, the debate between repetitive times *vs.* open, irreversible times deeply involves the psychoanalytic field. The concept of repetition, one of the focal points of psychoanalytic theory, is at the root of understanding the effects of psychic trauma, the neuroses, and other clinical presentations, as well as the configuration of the transference field. But if time is in essence repetitive, how do we explain the tendencies leading to psychic change?

As we said, our focus on the factor of time assumes that we take cognizance of the studies on time produced by other disciplines: physics, philosophy, and history. Thus, we learn that there is no one-dimensional interpretation of time. Some authors have accentuated a circular conception alluding to repetitive times: the eternal return. In this context, a crucial fact in the history of science is the displacement of the closed system of Newtonian mechanics, in which time is reversible, by the second principle of thermodynamics, where contingency and chance predominate, thereby making it a system of irreversible time. This aleatory character or disorder was analysed by the theory of chaos, stochastics, and the theory of probability. It is intimately linked to the notion of open systems, since they cannot be categorized only from within a given system.

In physics, Prigogine (1988) contributes this notion of irreversible time, based on the concept of dissipative structures. This opens the comprehension of time, exceeding, but not cancelling, the concept of incessant repetition. It also allows us to think about the subject, not on the basis of a psyche closed on itself, but open to others, which presupposes accepting the effects of time on the psychic subject.

In philosophy, Kierkegaard's ideas on "repetition with a difference", subsequently taken up by contemporary philosophers, add an essential variant for understanding processes of change.

In the psychoanalytic field, the significations assigned to the concept of time have powerful effects on: (a) the notion of history used in psychoanalysis: how do we categorize childhood history? What is the relation between the facts and the fantasies of each subject? Of this history, what is recovered and how? (b) The concept of repetition in action: how do we include the production of differences or, in other words, psychic change?

When we think about the relation between history and time, we know that the concept of history runs through Freud's works, with special emphasis placed on the structuring of the psyche and the vicissitudes of childhood history. Childhood traumata, sexuality and its traumatic condition, are all concepts referring to how and what may be recovered of childhood experiences. In Freud's works, developmental, linear, and progressive temporalities (for example, in his analyses of the psychosexual evolution of girls and boys up to the oedipal resolution and its desirable objectives) coexist with the concept of retroactive resignification that assigns signification *a posteriori* to a "previous traumatic fact". What is previous and what is subsequent enter into paradoxical relations. Linear chronology is dismantled and the material fact must be resignified.

For its part, the Kleinian School places the accent on progressive, developmental times: the passage from the paranoid–schizoid position into the depressive position as a treatment goal. Then the debate concerning the role of working through of the subject history *vs.* the "here and now" developed in post-Freudian psychoanalysis. It differentiated the mere reconstruction of events from their interpretation, which is always imbued with each subject's fantasies. On this point, we need to underscore that these fantasies are never arbitrary and always evidence relations, more or less distant, with a background of facts and events. These debates involve a merely developmental conception of psychic times—past, present and future—*vs.* the psychic work of constructing history and, hence, subjectivity.

As we said, Freud's works contribute a fundamental concept: retroactive resignification. The times of the trauma and its retroactive resignification provide a dimension that disassembles an exclusively developmental and linear conception of libidinal development concerning subjectivity. This also involves the introduction of a new signification, which is unframed from the original fact (while also preserving it).

In contemporary psychoanalysis, the concepts of time and history have become increasingly complex. It is evident that this trend offers us an opportunity to think about the inter-crossing of the different temporal dimensions imbuing the subject, an inevitable aspect of the analytic process. History is time past, but what is recovered is now the work of working through of the subject history, which carries the mark of both passing time and resignifying time. It is precisely the notion of history that gains different dimensions when a purely deterministic analysis is disassembled.

But there is also something new: something not recovered and yet unprecedented. For this reason, we need to include in these intersections the notion of event: something that, in becoming, escapes history and presents the prospect of thinking about the emergence of something novel on the psychic plane. This is truly contrary to the Platonic concept of essence. In this sense, the categories mentioned: event, contingency, and chance, are part of today's concepts of time. There are no foreseen times, predetermined exclusively in the past, in the field of the psychic fantasies. The concept of event (Badiou, 1990; Deleuze & Parnet, 1977) refers precisely to something unprecedented in the strictly deterministic sense, which is fundamental if we are to conceive of tendencies that lead to the production of psychic changes. In this reasoning, chance and event allow us to think about other dimensions of time that exceed those foreseen in classical linear chronologies or in the times of the eternal return.

These notions assuredly contribute fundamental elements to the conceptualization of transference as something more than eternal repetition and to the illumination of "something new that occurs" in the space and time of transference.

To summarize: we find continuities and breaks between subjective time and chronological time; between the inevitable decrepitude of the biological body with the passing of time and the timelessness of the unconscious; between linear, circular times and retroactive resignification; between facts, screen memories, memory, and the work of constructing history; between the times of repetition and the times of difference; between reversible and irreversible time; between the timelessness of the unconscious and the temporalities of the ego. The time arrow points towards an irreversible time, with no return, but coexisting with circular times and the times of repetition.

These plural, heterogeneous dimensions of time also enable us to think in terms of generating a prospective space of the time of becoming, of a desiring project or of anticipation, based on new versions of the past. In this context, we are interested in underscoring the time–space relation in the psychoanalytic field (psychoanalytic space, space of the session). Temporal and spatial relations have an intrinsic degree of relation. Space and time are part of the setting. This includes the times of the end of analysis, confronted with the timelessness of the unconscious. No less important is the way the passing of time works in the psychoanalyst, another variable worthy of analysis: the way it operates in the perspective of the termination of analysis and in relation to the analyst's own inevitable finitude.

The papers collected in this book illustrate these concepts with all the theoretical variations characterizing state-of-the-art psychoanalysis.

References

Badiou, A. (1990). *Manifiesto por la filosofía*. Buenos Aires: Nueva Visión.

Bellone, E. (1989). *I Nomi delTempo. La Seconda Rivoluzione Scientifica e ilMito della Freccia Temporale*. Turin: Bollati Boringhieri.

Deleuze, G., & Parnet, C. (1977). *Diálogos*. Valencia: Pre-Textos, 1980.

Hawking, S. (1988). *A Brief History of Time. From Big Bang to Black Holes*. New York: Bantam.

Hawking, S., & Penrose, R., (1996). *The Nature of Space and Time*. Princeton, NJ: Princeton University Press.

Prigogine, I. (1988). *Temps et Devenir*. Geneva: Patiño.

Vilenkin, A. (2006). *Many Worlds in One. The Search for Other Universes*. New York: Hill & Wang.

CHAPTER ONE

From the ignorance of time to the murder of time. From the murder of time to the misrecognition of temporality in psychoanalysis[1]

André Green

It is striking that the problem of time has been the source of far fewer discussions than themes relating to space. We have talked about the construction of analytic space (Viderman, 1970), of transitional space (Winnicott, 1953), but there is nothing analogous applying to time. It would seem that this theme has been avoided. Freud developed his ideas in a fragmentary and unsystematic way, as they appeared to him, and never brought together his diverse conceptions on time into a single presentation. Thus, he left us with a mosaic of temporal mechanisms without conceptual unification. After him, and profiting from this fact, analysts preferred, it seems, to circumvent the difficulty by not expressing an opinion on the unity to be identified in the diverse aspects described, instead of endeavouring to put the different facets of this concept into perspective. A tendency to return to the past—a regressive process—made analytic thinking return surreptitiously to a pre-psychoanalytic conception of time. In a more recent inspiration, it seems that the genetic approach, which for Freud was only one of the procedures for treating the subject of time, has progressively imposed itself in a predominant manner as the one that necessarily supplanted the others by eclipsing what stood out as specific to the theorization of the whole.

An immediate intuition makes us conceive of psychoanalysis as a fundamentally historical discipline, since it is a question, through it, of examining the consequences of a fixed or deviated evolution of development, linked to the vicissitudes of that which could not be integrated and has undergone multiple fates. The latter themselves influence the idea that one gets of the relations between history, its impasses, how the stages of its development are inscribed, its incapacity to resolve the conflicts that have arisen and their eventual subsequent return in forms that must be decoded if one is to understand how they are related to the experiences of the past that were unable to integrate them. All this is reproduced in each treatment and is related with both the vicissitudes and failures that are its conclusion. Attention has recently been directed towards the risks of the treatment, the uncertainties, and even the obstacles that stood in the way of recovery, as discussed by Freud in 1937; the relation of these diverse vicissitudes with the problems linked to temporality has been neglected.

It seems evident that the heart of the sphere of influence of psychoanalysis should be situated within the framework of the experience of the session, but this apparently obvious fact is counterbalanced by two remarks. First, numerous examples drawn from Freud's work show that the session does not have a monopoly on manifestations that bear witness to the organization and effects of the unconscious in relation to time. Let me just cite, to focus our ideas, the reactions arising from the contemplation or analysis of certain cultural works (*Moses*, Hamlet), or, again, certain experiences where the psyche has to accommodate psychic phenomena due to reorganizations of the past or the present reactivating in a new way the oldest wishes, etc. (*Leonardo Da Vinci and a Memory of his Childhood* [1910], "A disturbance of memory on the Acropolis" [1936a]).

To be sure, repression appears to be the principal mechanism responsible for a disturbance of memory that has preferred to reject into forgetting what it is unwilling to remember. However, this burying is scarcely passive. Fragments forming part of dream associations, co-opting and juxtaposing themselves by affinity, bear witness to a work of the un-conscious in conformity with what Freud called "the attraction by the pre-existing repressed".

The case of the dream, however demonstrative it might be, is more complex. The dream itself occurs outside the dreamer's consciousness, thus, by definition, outside the session, but it needs the work of the session in order to be interpreted. For Freud, the dream is a form of memory that survived repression thanks to disguises that render unrecognizable its relation with the past (the dream-work). Many other examples naturally lend support to this idea. More than any other manifestation, the transference refers to fragments of the past that could not be elaborated; it is clear that it goes beyond the framework of the session. One cannot expect less of a discipline whose aim was clearly the remembering of the forgotten years of childhood, the method being indistinguishable here from the therapeutic technique (Donnet, 2001), until Freud was obliged to recognize that the lifting of infantile amnesia, a desirable aim, was not always achievable, and that this modified the perspective of analytic work. But here, too, repression can concern details whose importance resides in their contiguity with the material to be repressed.

Once the repression has been identified, any psychic manifestation recognized as belonging to the return of the repressed is necessarily bound up with the past through the relations it allows us to surmise between the elements that reappear of that which had to be turned away from consciousness and distanced from it, and that which now demands to be heard in spite of the wish to reduce it to silence. This is not to limit the return of the repressed to that which was excluded from consciousness, for the condition of its return is to dress itself up in disguises that make it unrecognizable with regard to that which was condemned in the past and excluded from consciousness. To pass the censorship, it now assumes more accessible forms in order to escape a reinforcement of forgetting and to acquire once again a presence that obliges us to consider it as still actual.

The most demonstrative case of such work is the screen memory, which, like a collage, agglomerates recollections belonging to different periods of life. Freud even went as far as to assert that all of what is essential from infantile psychic life has been retained in the screen memory (Freud, 1917b, p. 148). Sometimes, the repression does not concern the lived experience, but one of its psychic elements.

Already in his study on screen memories in 1899, Freud notes the particularities of memory, rarely present in the form of a continuous chain of events. In fact it has been remodelled under the influence of a process: "conflict, repression, substitution involving a compromise" (Freud, 1899a, p. 308). In fact, one finds at the basis of what is remembered the displacement that permits the juxtaposition of phenomena belonging to different periods of childhood. Thus, events recollected dating from the post-pubertal period are contiguous with events of childhood. It is not only that the innocence of childhood permits their evocation, but rather that the contiguity with more clearly sexual memories suggests that they, too, were impregnated by a sexuality whose traces had disappeared and which are then surmised *après coup*. Likewise, important recollections coexist with indifferent recollections in order to mark their importance and to conceal their links with sexuality.

In short, the presence of the mnemic image is not a sufficient element for identifying the unconscious representation and for recognizing the significant element, sometimes constructed after former events have been recounted by the family circle. Besides, what seems important today did not have the same importance in the past, which is now re-emerging. What is more, associations show how a childhood impression can be revived. The falsification of memories that forbids the access of the original impression to consciousness through resistance serves the repression that dominates the experience and helps to substitute shocking and disagreeable impressions with other more innocuous ones.

In this first approach, Freud already draws on diverse types of temporalities, one of which, that linked to the development of the libido, is connected with a mode of evolution of a biological type, with the description of the successive phases of the predominance of the bodily zones of the libido. But this evolutive basis is already modified by experience that will mark more particularly certain stages, the fixations, and, subsequently, the tendency to return backwards towards the privileged fixations owing to the mechanism of regression. What needs to be noted here is the bidirectional tendency of the psyche, which is well illustrated by dreams.

However, in the course of this evolution, the memories of the epochs traversed seek to be put to advantage in an attempt to explain that which remains dissimulated by adults. The memories play a role in the construction of infantile sexual theories which will persist beyond elucidations into the real nature of the events concerned and will continue to be active in the adult unconscious: curiosity about the conception of children and the relations between the sexes, how pregnancy is accomplished.

In short, all memory is indicative; the rapport with that which had to be repressed remains the essential issue and can only be approached through the effects of contiguity, which invite us to surmise what the object of repression must have been and oblige us to consider a mode of temporality that is essentially different from consciousness (consciousness and memory are mutually exclusive, Freud says), or we rely on that which can be remembered according to the schemes of conscious memory. The most remarkable feature remains the absence of the wearing effects of the past in the manifestations that can be attached to the unconscious.

The study of the transference psychoneuroses would make it possible to confirm Freud's earlier ideas, the study on repression offering the most complete picture. Yet, Freud necessarily expected to discover other related forms which necessitated more nuances. For example, in his study on Schreber (Freud, 1911c), he distinguishes a form of repression that needs to be differentiated from the usual procedure encountered in the neuroses. In place of the general idea that what is supposed to be suppressed within (repression) comes back from without in hallucination, he substitutes another mechanism, wishing no doubt to radicalize the refusal of a psychotic nature characterized by this type of counter-investment: that which has been abolished within returns from without. So, it is not only what has been suppressed, but what has known, more than an annulment by consciousness, a veritable annihilation. This is what gave Lacan cause to describe the *Verwerfung* as different from the *Verdrängung*. This abolition could be understood as an erasure of the internal links constitutive of symbolization, which affects all the internal relations that call for different modes of interpretation than those of the neuroses, for the organization of the material bears the mark of this symbolic deficiency.

Other mechanisms would be described later on, as in the disavowal of fetishism (*Spaltung*), where Freud describes for the first time a defensive process that says simultaneously yes and no (*Verleugnung*) (yes, my mother has no penis; no, that cannot be true), accompanied by a displacement on to a secondary zone to replace the missing penis; this is the function of fetish-substitutes (suspender belts, stockings, etc.).

Last, the study of negation provides the opportunity for a new metapsychological construction, which shows Freud is eager to extend the process of negation beyond language, conceiving of a scale ranging from the earliest oral impulses expressed by all forms of rejection, to the later forms of repression, whereas affirmation manifests itself by the wish to take into oneself, to introject, to identify with. So, foreclosure, repression, disavowal, and negation form an ensemble whose common denominator is recourse to a mode of judgement by yes or by no which unites these different expressions and relates to what I have called "work of the negative" (Green, 1993). The repression of Oedipal impulses gives infantile sexuality its diphasic status.

On the other hand, Freud was to assume the existence of hypothetical factors that he introduced with the aim of categorizing the multiplicity of experiences, regrouping them according to the so-called primal fantasies. Not everyone accepted this hypothesis, but it is very difficult to see how the experiences are regrouped according to shared aims that escape consciousness. Such would be the role played by the fantasies of seduction, castration, and the primal scene, to which Freud was to add the taboo on omophagia. Freud gives them the function of being primitive fantasies. There was a wish to make the infantile sexual theories play this role, but their role is rather to reflect them. Lacan called them "key-signifiers".

The line followed up until then was to undergo a profound mutation with the introduction of the repetition compulsion. One might be tempted to see in it a complementary form of the ignorance of time defended by Freud at the level of unconscious phenomena. However, here the new situation introduced by Freud is, above all, designed to back up his idea that repetition stands in the way of the phenomenon of remembering, which he had hitherto considered as characterizing the essence of the therapeutic process. Repetition is not only a form of memory block. The compulsion to repeat takes the place of remembering. While the latter referred to a negation of temporality which made it possible to go beyond phenomena linked to the wearing effects of time, thereby resisting forgetting and allowing a desire to subsist which remains active, here it is a question of a form, which, through repetition, constitutes a denial of the temporal movement. It suggests a form of abolition close to foreclosure, a murder of time, the repetition exhausting itself through thwarting the effects whereby remembering makes it possible to infer that which is covered over and can no longer even make use of the opportunities opened up by displacement and disguise.

The modification that was introduced with the compulsion to repeat had been preceded by a change which was to lead to the cascade of changes marking the turning-point of 1920. Indeed, at this date, Freud unequivocally impugned the notion of the unconscious in the second chapter of *The Ego and the Id* (1923b). He contested there the importance of the role he had accorded to the unconscious representations which could participate in the memories that were themselves unconscious. He was now proposing a new form for the fundamental elements of the psyche: the instinctual impulse, which, let it be noted, contains no reference to representation and now constitutes the most primitive basis of psychic activity. It is evident that the theory of memory was thereby drastically changed. The "outside-time" now exists not as a system of traces anchored in the past and the unconscious, but in the form of a repressive dynamic traversing the psyche through the tensions of a raw libidinal erotism and of a destructivity which seeks to undo the links that have succeeded, not without difficulty, in forming themselves concerning the traces of this past.

However, even if the instinctual impulse is now considered as the material on which the construction of the psyche is built, according to Freud the dream continues to be a form of memory. It is as though the elaboration of the representative function made it possible to link up memories with unconscious representations that remain a vehicle for a disguised form of memory. Likewise, for Freud, it is no longer necessary to regard the transference as a transference of elements belonging to history and to the past. If it now depends on the compulsion to repeat, it induces us to make a deduction allowing us to infer the relations that have become inaccessible to consciousness between the past and the present. It is up to the analyst to render present these links between past and present; even if he considers that it is not always necessary to communicate them, he notes the discrete signs whose meaning is none the less clear, which he can postulate on this subject.

Nowadays, the interpretation of the transference will be direct, doing without justifications invoking its relation with the past.

In truth, though Freud could not immediately accept the consequences of his discoveries, these were already implicitly present as early as 1914 (the compulsion to repeat taking over from remembering, now set aside). The second theory of the psychic apparatus, placing the instinctual impulse at the basis of the psyche, was substituted for the unconscious representations hitherto conceived of as being directly linked to the unconscious. On the other hand, the id of the new topography was loaded with contents borrowed from phylogenesis, which were challenged by the majority of psychoanalysts.

Without any doubt, a radical turning had been taken, but it is impossible to evaluate the consequences without examining closely the originality of the conception of time, which Freud stacked up more than constructed, refraining at least from trying immediately to establish the coherence that constituted its originality. For the first of the realities from which we must begin is the observation of the heterogeneity of the psyche, with which different, not to say divergent, modalities of temporal exercise are connected. The first is that which sees time as an *a priori* of sensibility (Kant, from whom Freud differentiates himself), that is to say, as a component constituted outside conscious thought. However difficult it might be to give a clear definition of it, the triple categorization present–past–future can infiltrate surreptitiously a mode of thinking which, without having broken its ties with consciousness, is connected with the unconscious, as can be seen from the case of phantasy.

> We may say that it [a phantasy] hovers, as it were, between three times—the three moments of time which our ideation involves. Mental work is linked to some current impression, some provoking occasion in the present which has been able to arouse one of the subject's major wishes. From there it harks back to a memory of an earlier experience (usually an infantile one) in which this wish was fulfilled; and it now creates a situation relating to the future which represents a fulfilment of the wish. What it thus creates is a day-dream or phantasy, which carries about it traces of its origin from the occasion which provoked it and from the memory. Thus past, present and future are strung together, as it were, on the thread of the wish that runs through them. [Freud, 1908c, pp. 147–148]

Less interest should, no doubt, be accorded to the extensive tendency of that which conveys the phantasy—invading times other than the present in which it manifests itself—while being more attentive to timelessness owing to the fact that an excitation in the present bears witness to its living rapport with the non-present that surrounds its manifestation, marks the wishes related to the future, and unites the whole; "strung together on the thread of the wish" (*ibid.*, p. 148). We could conclude from this that it is in the nature of the wish to offer this thread, around which the temporal derivatives will be formed.

The idea of the thread lends itself perfectly to the idea of different conceptions of temporality revolving around one and the same axis. These conceptions already claim our attention in *The Interpretation of Dreams* (1900a), where the first formulations on the timelessness of the unconscious appear, the dream-elements including through association very different episodes and presenting the wish of the dreamer's life as fulfilled (optative). Freud repeats this insistence on the timelessness of the unconscious constantly throughout his work. On the other hand, the explanations he gives of it are very limited. Most of the time, Freud confines himself to stating that in the resurgences of unconscious material, the revived elements (memories, impressions) seem to be unaltered by time. If, today, it is possible to call into question and interpret in diverse ways the place of remembering in psychoanalysis, the freshness conserved of a buried and repressed mnemic evocation remains an indisputable fact, although there has been no occasion to evoke it before.

Freud lays much emphasis on what he terms the psychic nature of the material. What he means by that, although this usage raises problems, is that the facts to which he is referring must be distinguished from those that belong to perception, for "psychic" means here that their production has involved psychic work. One may ask whether raw instinctual material is apt for this elaboration. For a long time, Freud was to identify psychoanalytic treatment with a work of remembering, until he accepted the limitations of this aim; this movement corresponds to the growing importance of the transference and of its new hypothetical base: the instinctual impulses.

What is even more striking in the construction of Freudian temporality is the implicit confrontation of diverse orders of time. Thus, alongside remembering, one can speak of a psychobiological time, at work in the development of infantile sexuality. Another temporal organization is that of projections, which will play an organizing role in relation to future experiences of a psychic nature. This is the case of the infantile sexual theories or of the family romance. For the sexual theories are not only a recategorization of scattered elements, conscious or unconscious, belonging to the past; they also indicate a direction to be followed for the libido. These sexual theories will have an influence on the beliefs and orientations of sexual curiosity during subsequent reorganizations.

One of the most typical mechanisms of the organization of temporality according to psychoanalysis is the *après-coup* (*S.E.*, "deferred action"). Misappreciated by the English school, it is much in evidence in French psychoanalysis. It assumes that memory is not limited to a single impression, which is re-evoked in its entirety when remembering occurs. It postulates, on the contrary, that after this inaugural time, the memory does not remain frozen or fixed, but is susceptible in certain circumstances of emerging, owing to different but related associative contexts, showing a new vitality as a result of being linked up with the earlier contents that have become conscious again. In fact, the mobilization of the memory depends on certain links of an analogical order, which excite the primitive core and agglomerate around it, while prolonging and extending during this second time significations which had hitherto remained latent but which, thanks to the reactivation, can emerge and enrich what has already been signified in relation to perception, to the event, or to the early memory.

Freud had already become aware of this in the "Project" (1895), in one of the first attempts to describe the specificity of mnemic and temporal functioning in the "proton pseudos", the first hysterical lie. It was presented there not as resulting from the evocation of a memory, but as resulting from an association to associations, which was in no case present in the first volley of associations of the memory. This indirect resurrection pleaded in favour of a persistence in the unconscious between the consequences of the first mnemic evocation and the appearance of a later chain of events which, because it allowed more light to be shed on the symptom, addressed it and added to the meaning of the first one.

Freud was to take these ideas up again in a work of maturity much later on, in the analysis of the Wolf Man (1918b). He had already noticed that the remembering of the trauma, more than its immediate effect, was full of consequences for the psyche. As he was led to reconsider the temporal localization of the trauma (in this instance, the primal scene), he argued in favour of its manifestations of resonance: dreams related to the observation of copulation, displacement on to the observed copulations between animals. He then posed the hypothesis of a deferred action, remaining subject to forgetting between the early trauma (at the age of one and a half) and the subsequent reactivation (during the dream, at the age of four). He writes, "(I purposely avoid the word recollection)" (*ibid.*, p. 44), to clearly distinguish it from a mnemic evocation, no doubt due to a more important charge linked to the connection with the unconscious.

It is clear, then, that the early impressions are situated at the heart of an associative network constituted from them and susceptible to having a retroactive effect on its source, which is still alive. He was to call this effect due to the latest periods of time "deferred action" (*après-coup*) (*ibid.*). This description raises the question as to whether it is not inappropriate to speak about the events in connection with the deferred effects in terms of memories or of co-optation subject to the pressure of the unconscious.

How are we to designate this latency, which is capable of coming to the mind "as if it were a memory"? The idea of a mnemic latency could tally with cases where the early impression and the deferred effects remain susceptible to activation in the unconscious. The cut between the two series of events could be said to play the role of a censor that allows itself to be taken advantage of. What is involved here is the representability of the pair, primitive excitation, and its later effects. This occurs when the early events are neither understood nor conceivable, even in a symbolized form. This is what is often observed in borderline personalities, who generally lack the capacity for representation that is capable of suggesting an evocation *après-coup*. They conserve, however, a latent capacity for recognizing analogies which make it possible to evoke a first form of a matrix of memory. However, there are cases where the early impression is no longer capable of being "submitted to reacquisition" (Freud) and thus cannot play the role of a form of appeal for the constitution of a pair whose significance is revealed by the relationship between its parts. This is what we often encounter. But it will be a precarious form, with little possibility of connecting itself up with new associations that would make it possible to link them up with an earlier trace.

Freud excludes the idea that the deferred effects are mere products of the imagination. He wishes to insist on their participation in the foundations of the psyche, that is to say, on the fact that they belong to wishes that have acquired the value of postulates for subsequent beliefs, assuming the function of a substitutive truth. Here, construction already asserts its rights and aims to restore a distant form of reality that Freud calls instinctive impulse, in order to distinguish it from the memory. It carries with it the form of conviction that we recognize in psychic reality.

Equally, there appear ideas of conceptual value that have been the subject of discussions, such as the idea of phylogenetic schemes, which, according to Freud, operate a categorization of experiences, reordering them, giving importance to some more than others, as if to fulfil a pre-established programme in which everything must find its place in a certain order, less temporal than categorial, that is to say, according to a prioritization of categories. Freud even goes as far as to consider that certain fantasies might increase the role and significance of events that have not marked the history of the subject sufficiently. If certain types of experiences only have a rather weak reality, it is the reference to these schemes that gives them the force that has been lacking in such an individual history, in such a way as to gather up the diversity of experiences subject to the effect of "primal fantasies", the gathering together of which gives coherence to the erring ways of infantile sexuality. Such are the roles of the seduction, castration, and primal scene fantasies, to which will be added subsequently that connected with omophagia, in relation with cannibalistic tendencies.

A long reflection, which would exceed the limits of this study, would show what might be designated as the implicit logic of these categorizers. All the manifestations I have cited remain part of the general framework of an extra-temporal temporality and concern origins, as Laplanche and Pontalis have pointed out (1973). Their unconscious nature or the total absence of any reference to the logic of consciousness throws light on what animates them. All this was to culminate for Freud in the Oedipal organization, the crowning achievement of infantile sexuality.

With the intervention of the compulsion to repeat, yet another variety of time takes its place in the Freudian theory; a variety of time that is not only unconscious (unaware of time), but the fruit of a compulsion that contests the seminal role of the pleasure principle. Here, it is not only that which resists time and serves it that imposes itself, but that which imposes itself on conscious temporality in order to give precedence there to the aim of re-establishing an earlier state.

It is very clear that Freud, in so doing, has changed direction. He argues less for a sort of eternal youth of the unconscious than for a rebellious, inexorable temporality, refractory to the expansions of the libido—even when the latter refers to events that are long past—destined not only to survive the wearing effects of time, but to drag the latter towards a deadly slope where nothing is held back any longer, blocking all evolution, freezing the general orientations of a libido that had hitherto authorized successive agglomerations, destined to founder in the monotony of a rigid vitality, denying itself the contribution and enrichments of that which is new and capable of casting a fresh view on the past, thereby suggesting that it can be seen differently.

Today, many analysts contest that a real compulsion to repeat exists; their sensitivity to detecting differences behind the returns of the same makes them prefer another solution than that offered by the compulsion to repeat. Yet, it is present in hidden forms under the material and justifies the denomination of the compulsion to repeat. Nevertheless, psychoanalytic literature has not ceased, since the date when the concept was put forward, to concern itself with this problem, which has been envisaged in turn by Freudian and Kleinian analysts, ego-psychologists, or analysts of other orientations. All these different currents have thrown light on the rigidity of the psychic organizations in which the compulsion to repeat prevails, without any new conception imposing itself.

It is clear that, in invoking the compulsion to repeat, what we have in mind is not a repetition of the identical, as this is not what we generally encounter. However, it is equally clear that the differences observed within the framework of repetition cannot persuade us that these forms of psychic life can profit from the enrichments of what is new and from the opportunities life provides for the satisfaction of their wishes. This, indeed, is why we continue to attribute value to Freud's observations. In any case, their mobilization will depend on the support drawn from the transference.

However, we can go further than Freud and underline a sterilizing, asphyxiating action, owing to a closure to the contributions of what is new, in which temporality brings less the choice of possibilities exhumed in relation to the past than it gives the impression of acting, not in the manner of the unconscious outside-time, but in the manner of an anti-time, that is to say, a murder of time which the patient has the capacity to stop. It is this difference that it is important for us to elucidate, the triumph of the outside-time in favour of the survival of desire, opposed to getting bogged down in a repetitive compulsion which is an enemy of life and of what is new and still to come.

It is indeed remarkable that if the outside-time of the unconscious is primarily perceivable through contexts of an evolving transference relating to fixations of the libido, momentarily immobilized by the fixations, the deadly forms of repetition are generally encountered at the heart of structures dominated by masochism, aggressivity, and the unconscious need for self-punishment, which seem to be frozen in their activity of unbinding and in their forms of iterative destructivity which cannot have any other aim than that of reproducing themselves. More generally, we will find ourselves faced with the effects of the compulsion to repeat when, by default, intermediary formations (fantasies, representative constructions, structural antagonisms) are lacking in the material; this is often observed in borderline cases where associative modes are found that short circuit thought or prefer to deny it in an omnipotent attitude (Bion, 1962).

Hence, the necessity for a new temporality, one that is no longer based on remembering that is as aleatory as it is undoubtedly mythical, but on the response that can come from the analyst faced with the blocking of temporalized significations. I want to speak here of the activity of construction, which, though it is more aleatory, offers a chance for elaboration, which from now on lies in the hands of the analysand alone, and also of a self-elaborative and self-critical activity of the analyst, inviting him to make use of all the elements of the material permitting the organization of a hypothetical picture, whose dynamic and transformational possibilities will allow a processual temporality to be set in motion again.

Freud himself was to arrive at this conclusion, shortly after having given way to the pessimism of "Analysis terminable and interminable" (1937c). But, by the same stroke, he also found again the unity of psychoanalysis through a broader conception of reminiscence, reminiscence that now uses the imperatives of repetition, but that allows us to make use of it for constructive ends; that is, through the constant correction of his initial attempts to find meaning by others, no doubt more audacious still, owing to the increased intelligibility that their analysis permits.

And, from a point of view that is increasingly removed from realism, it perhaps also gives a chance to what is implied by the unconscious process, and to what we can notice about a conception that is increasingly removed from habitual modes of thinking (Freud, 1940), soliciting, through interpretation, the creativity which animates every psyche—especially during the encounter of two interconnected, meshing, and mutually complementary psyches—so that each of them gains its liberty while recognizing the contribution of the other.

Today, if the hypothesis of the death drive remains poorly accepted, it would be contrary to the facts to deny the blocking effect on the analytic process of the mechanisms of repetition, which needs to be linked up with the attempts to prevent thought from developing or from extending the context that permits a non-conscious causality to be apprehended. The precocity of traumas prior to the acquisition of language prevents them from being kept in reserve in the form of memories.

And yet, this was the reason why Freud was ultimately obliged to recognize the limitations of infantile amnesia and to accept the recourse to construction. However, at the basis of the latter, there subsist effects of reminiscence that cannot be considered as memory but that belong to forms of fragmentary remembering that seek to be reunited in larger ensembles than the analyst's construction is able to identify.

In short, even that which claimed to exclude itself from any form of attestable temporality serves as material not only for evoking the past, but also for the construction of a future which conserves with this past relations of consonance that not only require us to understand what was rejected from consciousness, but also induce us, through construction, to have access to the reasons why they were excluded from consciousness. This is where the work of elaboration draws—differently according to the theories—on the hypothetical constructions of the genesis of the psyche.

Above all, the bidirectional approach of analysis must be preserved: one takes a retrogressive path, allowing us to anticipate the beyond of a current approach, while the other enables us to get a better idea of the past through fantasies of the future which seek to bring the past back to life, making it pass for something new.

An organization of temporality favourable to psychic structuring does not consist, then, in the conservation or availability of an important mass of memories, but in a supple and mobilizable network of a reserve outside time, which, according to circumstances, can be activated as psychic materials that permit us to surmise the main lines along which it may take shape through what we call "intermediary formations": dreams, fantasies, bungled actions, and any other manifestations of the unconscious rendered permeable owing to the work of the preconscious. "Permeable" means here "which authorize their deduction". In other words, it is the existence of a restored signification concerning psychic events that can be linked up with a historical truth.

The historical truth is not the truth according to a history constructed from the outside but, as Freud had already indicated, the truth that prevailed at the moment when beliefs were formed in relation with the unconscious. This accessibility depends on the defences employed not being too rigid, allowing for a certain free play that makes it possible for a system that finds itself too closed off in the unconscious to pass by another route that opens up. But, let me repeat, it is only through direct communication of the memories of events that one can have any chance of making constructions that have a value of conviction for the one for whom they are destined.

Modern psychoanalysis has turned away from the search for the multiple paths necessary for temporal construction, perhaps owing to deceptions resulting from the speculative exercise which endeavoured to answer enigmas. To replace it, it only found, in my view, impoverishing solutions, such as the technique of the "here and now", which comprised no fewer hazardous speculations by relating everything to a present arising from the thought of the analyst alone, no less debatable in the forms that it was supposed to take.

Another solution consisted in falling back on a very narrow genetic point of view, often inspired by observations of children that are incapable of embracing the internal movements animating the psyche "of the one who is being observed".

However, from an opposite point of view, through the contributions of contemporary psychoanalysis, one can highlight factors favouring the disorganizations of temporality, or, failing this, factors impeding its integration. I will simply cite them here. They either hinder the unconscious–preconscious communication or, more deeply still, seem to thwart the activity of binding at the unconscious level:

- early traumatic experiences occurring before the appearance of language, unelaborable in terms of remembering, particularly under the auspices of terror, which could even include psychic catastrophes and massive bereavements;
- the absence or deficiency of the transitional space;
- the failure of reparation;
- the excessive degree of projective identification or expulsion, which eventually becomes an obstacle to any conservation that can be remembered;
- ego-distortions of the type of disavowal and foreclosure and, of course, the effects that we have already envisaged of the compulsion to repeat.

In conclusion, the normal structure of temporality is that of an exploded time (*temps éclaté*), diffracting those past experiences that inevitably succumb to the work of the negative because they are linked to prohibitions that no longer permit them anything but an unconscious survival unavailable to memory or, more profoundly, removed from any form of consciousness or representability except in their raw expression. In short, the representative sphere is curtailed in favour of the unrepresentable.

This leads us to postulate a more or less open ensemble in which a polychrony of psychic life is expressed, reflecting an exploded time that is sometimes susceptible to activating some of the aspects of its functioning. In any case, a bidirectional approach to time must be preserved, the existence of which we observe each night following each day, where the day-time work is undone so that another may take its place: the dream-work. Thanks to the activity of dreaming, life becomes bearable by conserving in us a portion of hope in the form of illusions to which our being owes the capacity to bear the inevitable disillusionments to come. It might be said, as a concluding remark, that the whole work of psychoanalysis is one of recognizing that which forms the foundations of our identity.

Notes

1. Translated from the French by Andrew Weller, Paris.

References

Bion, W. R. (1962). A theory of thinking. *International Journal of Psychanalysis, 40*: 306–310.

Donnet, J.-L. (2001). De la règle fondamentale à la situation analysante. *Revue Française de Psychanalysis, 65*(1): 243–258.

Freud, S. (1895). *Project for a Scientific Psychology. S.E., 1*: 281–397. London: Hogarth.

Freud, S. (1899a). Screen memories. *S.E., 3*: 301–322. London: Hogarth.

Freud, S. (1900a). *The Interpretation of Dreams. S.E., 4 & 5*: 1–621. London: Hogarth.

Freud, S. (1908c). Creative writers and day-dreaming. *S.E., 9*: 143–153. London: Hogarth.

Freud, S. (1910c). *Leonardo Da Vinci and a Memory of his Childhood. S.E., 11*: 63–137. London: Hogarth.

Freud, S. (1911c). Psycho-analytic notes on an autobiographical account of a case of paranoia (dementia paranoides). *S.E., XII*: 1–82. London: Hogarth.

Freud, S. (1917b). A childhood recollection from *Dichtung and Wahrheit. S.E., 17*: 145–156. London: Hogarth.

Freud, S. (1918b). From the history of an infantile neurosis. *S.E., 17*: 1–122.

Freud, S. (1923b). *The Ego and the Id. S.E., 19*: 3–66. London: Hogarth.

Freud, S. (1936a). A disturbance of memory on the Acropolis. *S.E., 22*: 239–250. London: Hogarth.

Freud, S. (1937c). Analysis terminable and interminable. *S.E., 23*: 209–254. London: Hogarth.

Freud, S. (1937d). Constructions in analysis. *S.E., 23*: 255–269. London: Hogarth.

Freud, S. (1940 [1938]). *An Outline of Psycho-Analysis. S.E., 23*: 141–207. London: Hogarth.

Green, A. (1993). *The Work of the Negative*. A. Weller (Trans.). London: Free Association, 1999.

Laplanche, J., & Pontalis, J. B. (1973). The *Language of Psychoanalysis*. D. Nicholson Smith (Trans.). London: Hogarth.

Viderman, S. (1970). *La construction de l'espace analytique*. Paris: Gallimard, 1982.

Winnicott, D. (1953). Transitional space and transitional phenomena. In: *Playing and Reality* (pp. 1–25). London: Routledge, 1971.

CHAPTER TWO

A problem with Freud's idea of the timelessness of the unconscious

Charles Hanly

> "What then is time? If no one asks me, I know: if I wish to explain it to one that asks, I know not"
>
> (Augustine, *Confessions*)

> "Nature's bequest gives nothing, but doth lend"
>
> (Shakespeare, *Sonnet* 4)

It is natural for human beings to resent time and to phantasize about timelessness. We are biologically driven to seek to preserve ourselves. Even Freud's death instinct (1920g) was thought by him to be bio-chemically committed to preserving individual life from accidental death in order to bring about the end of it in its own way and time, like an avenger who wants to do the murderous deed himself. But even in the absence of a death instinct, the instinctual aim of self-preservation cannot be satisfied, because we depend on our perishable physical bodies for the existence we seek to preserve. Accordingly, we phantasize about timelessness both individually and collectively. These phantasies give rise to credence in seductive ideas such as Plato's imperishable soul or Descartes's mental substance linked to, but ontologically independent of, the body. In addition to the unavoidable neurotic anxieties consequent upon our individual development and the relations that are necessary to it, among the realistic anxieties caused by the dangers of our circumstances and the frailty of our ability to deal with them, there is anxiety about death. This anxiety is ontological; it is an anxiety about being a finite individual creature biologically destined to annihilation even if the species persists much longer. We resent being the finite, temporal, somewhat rational animals we are born to be and we seek to patch up this flaw in our being with ideas: among others, of an immortal soul. Anxiety about death is to natural reality what anxiety about our ownership of the Oedipus complex is to psychic reality.

Before the revolutionary discoveries of Copernicus and Galileo, mankind inhabited a cosmos, deeply rooted in belief and our perceptual experience of nature divided into sub-lunar time and celestial timelessness. This experience of time was taken for granted in the cosmologies constructed by Plato, Aristotle, and Ptolemy. Motion in the sub-lunar terrestrial region was thought to be rectilinear, having a beginning and an end and, therefore, finite and temporal, whereas celestial motion was thought to be circular, without beginning or end and, hence, continuous, unchanging, and timeless. Thus it was that, when our ancient and medieval ancestors of the western world gazed into the night sky, they were looking at, as they thought, "the moving image of eternity". Sun, moon, planets, and stars unchanging in themselves, as they believed, change their location relative to the earth in space by means of a changeless, unending motion. The discoveries of Copernicus and Galileo brought an end to this way of thinking about nature, even though their discoveries did not change the visual appearance of nature. And, we still hunger after evidence of the unchanging and timeless in our otherwise universally mutable universe. The two questions I propose to consider are: does the timeless make its appearance in psychoanalysis, and, if so, where, how, and to what effect? For this purpose, I shall limit my discussion to Freud. I shall argue that the references by Freud (1896a, 1900a, 1918b, 1920g, 1933a, 1939a) to the timelessness of the unconscious are empirically mistaken and hearken back to the narcissistic longing for indications of the immutable and timeless in human nature as found, for example, so passionately, ingeniously, and brilliantly, but illogically, argued by Plato (*Phaedo*) on behalf of his belief in the immortality of the soul.

Before turning to these questions, it is useful to briefly discuss the subjective experience of time. This phenomenon has been extensively discussed in the literature. The argument is that the subjective experience of time should not be confused with either the subjectivity of time or the physical relativity of time. I have set out the first part of the argument elsewhere (Hanly, 1997) and shall draw extensively from it here, while adding the second part of the argument relating to the relativity of time.

> In depression, time seems to be without a future; interest in the future is swallowed up in the misery of the encumbered present in which nothing changes or can be changed. Children typically experience time passing slowly—an experience, possibly formed by the longing to be adult in the fond belief that adults are able to do as they please. An afternoon waiting for a satisfaction can assume the aspect of an endless time. Youth, at its best, experiences time manically. The future appears to offer endless possibilities. In youth, the years of life are numbered according to the logic of the small child who can count to seven and identify classes of seven things correctly but who also uses "seven" to connote any number of anything greater than six, however great. Middle age, brings us, at last, close enough to average life expectancy to be able to count our years less delusionally. This realism is aided by encounters with limitations, among them the passage of time, accidents and the reality of adult helplessness in a largely satisfying life giving rise to the experience that time has accelerated as the seasons rush by. Until, subjective time slows down again with the declining strengths, increasing vulnerabilities and not infrequent depression of old age. [*ibid.* p. 8]

Analysts (Arlow, 1984; Orgel, 1965) have noted and analysed fluctuations in the subjective sense of time.

These and other fluctuations in the subjective sense of time have psychic reality and unconscious as well as conscious meaning for the individual. For this reason, they are of great interest to the analyst. However, these fluctuations, which have everything to do with the life of the individual, have nothing to do with the unidirectional pace of natural time. Subjective fluctuations of time are just that; they do not correspond with anything real in time itself. It would be a narcissistic delusion to suppose that a subjective slowing down of our *sense* of the passage of time could in the least affect the *passage* of time. Kant's (1781) idea of time as a pure form of sense experience goes so far as to render time subjective (i.e., psychological and not physical in nature), which is delusional enough, but not so delusional as to assume that human moods have any influence on either the rate or direction of time's unidirectional passage. Even so, Kant's idea narcissistically exaggerates human subjectivity in order to deny the majestic indifference of time to which our lives must submit. As Hawking (1988) points out, the unidirectionality of the psychological sense of time (we remember the past but not the future) derives from the brain's physical obedience to the laws of thermodynamics. The subjective fluctuations of the sense of time have nothing to do with the physical relativity of the real time of nature. The subjective fluctuations in the sense of time are relative to the individual's moods and unconscious phantasies; in nature, time is relative to the velocity of matter in motion.

What, then, are we to make of Freud's often repeated idea of the timelessness of the unconscious? Apart from some earlier hints, Freud (1900a) first states ". . . it is a prominent feature of unconscious processes that they are indestructible. In the unconscious nothing can be brought to an end, nothing is past or forgotten" (p. 577), and later (Freud, 1896a, in a footnote added in 1907), "In the case of *repressed* memory-traces it can be demonstrated that they undergo no alteration even in the course of the longest period of time. The unconscious is quite timeless" (pp. 274–275). And again Freud (1915e) wrote,

> The processes of the system Ucs. are timeless; i.e. they are not ordered temporally, are not altered by the passage of time; they have no reference to time at all. Reference to time is bound up, once again, with the work of the system Cs. [p. 187]

Freud (1918b) counselled patience with the timeless unconscious of the Wolf Man, and, finally, Freud (1920g) tells us once more, after referring to the Kantian notion of time and space as "necessary forms of thought" (to be correct, Freud should have said "necessary forms of perceptual experience"),

> We have learnt that unconscious mental processes are in themselves "timeless". This means in the first place that they are not ordered temporally, that time does not change them in any way and that the idea of time cannot apply to them. [p. 28]

The crucial summary text is Freud (1933a),

> There is nothing in the id that could be compared with negation; and we perceive with surprise an exception to the philosophical theorem that space and time are necessary forms of our mental acts. There is nothing in the id that corresponds to the idea of time; there is no recognition of the passage of time and—a thing that is most remarkable and awaits consideration in philosophical thought—no alteration in its mental process is produced by the passage of time. Wishful impulses which have never passed beyond the id, but impressions, too, which have been sunk into the id by repression, are virtually immortal; after the passage of decades they behave as though they had just occurred. They can only be recognized as belonging to the past, can only lose their importance and be deprived of their cathexis of energy, when they have been made conscious by the work of analysis, and it is on this that the therapeutic effect of analyst treatment rests to no small extent. [p. 74]

Let me first take up an issue of philosophical scholarship. Kant's (1781) theory is that space and time are necessary forms of perception and, in particular, that time is an *a priori* (prior to experience, non-empirical), pure (unmodifiable by experience) form of inner sense, grounded in the cognitive activities of the mind. After 1929, with the completion of her analysis with Freud, Marie Bonaparte, who was keenly interested in Kant's ideas, rekindled Freud's interest, although his thinking about Kant's idea of time dates back to 1907. Here Freud (1933a) offers a refutation of the Kantian notion of time as a pure, *a priori* form of inner-sense awareness. If unconscious mental processes and contents are timeless, then time cannot be a necessary "form of our mental acts". However, Freud's refutation is unsound.

Kant did not claim that mental acts are necessarily temporal. On the contrary, Kant believed that the moral activity of the will is a manifestation of a timeless noumenal self. Kant's idea was rather that time is a necessary condition for experiencing internal mental activity, including being aware of our perceptions of the activity of persons and thing external to the mind. Freud had forgotten his earlier position, consistent with Kant's idea of time, that ". . . time is bound up with the work of the system Cs"; i.e., with the experiential, cognitive work of conscious sensory activities. Kant thought that it is the perceptual system that systematically assigns temporal relations of simultaneity and succession to internal and external events. According to Kant, these events are not intrinsically temporal; they acquire their temporality from being subjected to a law (a pure, *a priori* form) imposed by the mind as a necessary condition for experiencing any thing, event, or activity. Unconscious mental processes are rendered temporal by being made conscious. Consequently, Freud's assertion that unconscious processes do not recognize time does not refute Kant's notion of time. These processes are not conscious; they are not experienced, although their derivatives are, and their derivatives are temporal. When, for example, a memory becomes conscious, it is immediately located in time as being before, after, or simultaneous with other remembered events in the individual's life, even though the exact location of an event relative to others may not be clear or even if it is, in fact, given a mistaken location in the temporal sequence of life events.

Although Freud's argument does not refute the Kantian notion of time, Freud's thought is profoundly incompatible with Kant's subjective idealism. I believe that Freud was correct in rejecting Kant's concept of time, although mistaken in the grounds he chose to justify the rejection. A more powerful argument was available to Freud at the time in Einstein's physics, in which time is more plausibly considered to be a dynamic property of the physical universe, relative to proximity to mass and to velocity (Hawking, 1988) and, therefore, not in the least reducible, as Kant thought, to a pure form of inner perception.

Unconscious processes are made up of contents: memories, thinking by means of images, wishes, aversions, and fears, and their combination into phantasies which have agency (motivational efficacy) on account of the sexual and aggressive drives that invest them. Among these contents, Freud mistakenly included ideational archaic inheritances (Freud, 1939a) which, if they existed, would have a transgenerational genetic immortality dating from the origins of mankind. But this Lamarckian hypothesis is scientifically untenable. In addition, genetic immortality is not timeless, since it will end with the extinction of our species (Hanly, 1997). Moreover, as powerful as the hypothesis would be as an explanation of the universality of the Oedipus complex, if it were true, it is not at all essential to psychoanalytic theory, since the inevitability of the Oedipus complex can be satisfactorily explained without it.

The father whom a small boy wishes to have out of the way is not the echo (memory imago) of the primitive father of the primal horde; he is the real father or his surrogate. But what of this phantasy father, the father who is symbolized by a dangerous monster in an unconscious phantasy against whom, in a further phantasy elaboration, he pits himself in a test of courage and strength? These symbolic substitutes of the father are not in the real world, they inhabit the worlds of imagination, although certain symbolic equivalents are found in the worlds of movies, comic strips, and literature. What of the phantasy mother represented by witches? What of the princess who symbolizes the small girl who is rescued by Prince Charming from the toils of her cruel stepmother? Are these unconscious contents and the processes that give rise to them atemporal and immune to mutability? What is the nature of their "virtual immortality"? Has Freud, having accepted the loss of cosmic timelessness, introduced into his idea of the psyche a repetition of the ancient time–timelessness dualism of the description with which we began?

The images that form these fantasies are surely temporal, whether they are conscious or unconscious. It is their intentional objects that seem to escape time. But these objects, too, can only do so by achieving an existence in the world of imagination which itself depends for its existence on the psychic life of persons or the world of cultural artefacts where they are once more temporal, however intensely invested with narcissism they may be. Freud would not be alone in finding such mysterious elements as archaic residues in the psyche for they are of a family with Platonic memory, Aristotelian *nous*, and Kantian noumenon. Let us explore the essential elements of Freud's meaning by means of a typical example of repressed unconscious elements and processes.

A common enough calamity of childhood is the birth of a sibling, especially when a child is under three years of age and remains very close to the special pleasures of infancy: the oral pleasures of breast feeding and the narcissistic pleasure of being the baby and of feeling herself to be the unique object of her mother's love. When a sibling arrives, the newborn appears to the child to have taken from her the love of the mother that vouchsafes these pleasures, pleasures that have nourished her well being and upon which her survival depends. The child's profound anaclitic attachment to her mother is disturbed and might be traumatically disturbed. Even the efforts at separation and autonomy of a two-year-old are grounded in the mother's love, which, when a new baby arrives, can seem to have been lost to the rival. It is not surprising, under these circumstances, that a two-year-old would develop an ambivalent attitude of love and hate toward the baby. The child's destructive hate thrusts her into a painful dilemma. The child hates the baby because she recognizes that the mother loves the baby, as she so recently loved her; she fears that the baby will take away the mother's needed love. Thus, the child becomes anxious lest the mother, who loves the new baby, will hate her for hating the baby. An internal conflict is generated. One way of resolving the conflict is for the child psychologically to deny her hatred for the baby by intensifying her affection for, and identification with, the baby. This reaction formation has the effect of repressing the memory of the hostile, destructive feelings for the sibling and all the phantasy and real experiences to which they had given rise. Henceforth, the jealous child is only able to experience affectionate feelings toward her sibling. A precarious relief from the jealousy will have been achieved. Relations with the baby will become more peaceful. The child might well become mother's helper in caring for the baby. The child's destructive hostility will no longer be experienced as such. The pleasure that the parents take in this development, rewards it without, by itself, mitigating the denied hostility. For the child, it is as though the painful episode with its tantrums, sulks, regression to thumb sucking, difficulties with sleep, bungled efforts to get rid of the intruder, had never happened, although versions of these symptoms might well continue, including a subtle artificial sentimentality, superficial complacency, and lassitude. We have been imagining the life of a girl. Gender difference does not immunize boys from the same experiences and consequent difficulties in life.

A personality is formed that does not take this important episode into account and which experiences as alien within itself the seemingly haphazard manifestations of destructive aggression in phantasies, dreams, play, and peer relations. If the latency child finds that the parents take pleasure in some activity of the younger sibling, the older one might shy away from engaging in it because of vague anxiety about being in competition, even though it is an activity that the older child enjoys. When, in adulthood, he marries and has a child of his own, he may find himself beset with obsessional thoughts about being guilty of some terrible crime. Although he knows that he has not committed a crime in reality, he feels himself to be a criminal. He is horrified by thoughts that he might push the carriage with his infant son in it into the path of a car when he is taking his newborn for a ride in the carriage. This father might find himself becoming overtly upset when an older child behaves aggressively towards a younger child, as though he feared that some terrible violence was about to erupt. Children, he believes, are supposed to only love and care for younger siblings, as he proudly remembers having cared for his younger sibling.

In this history, we have evidence of the continuing activity of repressed memories, impulses, and feelings from childhood. It was this sort of evidence that Freud had in mind when he spoke of the "virtual immortality" of repressed memories. It is evident, from these facts, that Freud used the words "timeless" and "immortal" poetically in order to express a narcissistic wish or a stoical irony. The essential psychoanalytic meaning of the terms refers us to the continuity and durability of unconscious organizations of the kind we have been considering. Second, it indicates the failure of repression and other defences to obliterate memories and phantasies or to extinguish libidinal and aggressive investments in them and, hence, their capacity to cause dreams, symptoms, and parapraxes, to interfere with reality testing, to distort character, to inhibit the ego, and to disturb object relations long after the original trauma had occurred. The third essential meaning is that unconscious processes are indifferent to time. Repression has caused the memories and phantasies, as it were, "to stand still" in time by remaining unchanged in their substance and action, although not in their derivatives. Finally, there is a fourth meaning, which is an implication of the first three meanings: time does not heal unconscious conflictual organizations of this kind. The wish to get rid of the unwanted sibling remains contemporary; it continues to seek satisfaction in the here and now, which, on account of repression, can only be substitutive and disguised; it refuses, as it were, its location in the past. The young father in our example feared that he could harm his son because he had once wanted to harm his mother's baby, without being able to remember it. His current sense of guilty apprehension may be corrected by the reflection "But I love my baby and far from harming it, I feel protective of it", giving rise to a search for something done in the past which goes nowhere and leaves the troubled parent anxiously aware that it is his child who seems to be the object of his criminal wish unbelievable as that also seems to be. The repression renders him unable to find out from whence his horrifying thought comes. This confusion is only increased by his having no such animus against the sibling as he/she now is, an adult like himself.

The father in our example feared that he would harm his infant son because he had once wanted to harm his mother's baby. In this way, the originating repressed memory has the appearance of an "eternal now". However, this appearance is an illusion of consciousness in the way in which the impression of uncaused choice is an illusion of consciousness. In using the words "timeless" and "immortal", Freud is drawing on an ancient identification of the timeless with the unchanging found in the Parmenidean (Proem) idea of being, the Platonic (Republic) idea of forms, the Aristotelian (Metaphysics) idea of completely actualized matter, the Ptolemaic notion of celestial circular motion, the Cartesian (1641) idea of mental substance and the Kantian (1781) idea of noumenon.

However, the four intrinsic meanings of timelessness in the psychoanalytic theory of unconscious contents and processes bear none of the metaphysical ontological freight of these philosophical ideas of the timeless. In this respect, Freud's assertion is mistaken. Unconscious processes are not themselves timeless, contrary to Freud's (1896a, 1920g) assertions that they are intrinsically timeless and that time does not apply to them. Unconscious contents originate with developmental or object relational calamities (usually both intertwined); they have a beginning; they will have a variable history according to their influence in the life of the individual and of the individual's relational life on them; and they will end with death. In themselves, unconscious processes are no less temporal than conscious psychic processes. It is just that they function differently, and one of the differences is that they are not, once established, easily subject to the influences of experience, development, and relations. Nor are they, on account of the amnesia of repression, located in the sequence of life events in which the individual finds his/her identity. But they are not at all immune to change in the sense in which, for example, Kantian noumena or Platonic forms are. Moreover, they can be changed either by good fortune in life of an appropriate kind (the inheritance of wealth, in itself, will not do, luck in love might) or by psychoanalysis. What cannot be changed by time, what has not been changed by life, still can be changed by psychoanalysis, as Freud (1920g) recognized.

It is this fact that renders inconsistent Freud's attribution of timelessness to the unconscious. Trauma can cease to cause psychopathology. Memories can cease to act as though they were current rather than past experiences. When they do, they take their place in the temporal sequence of the individual's life experience that they always actually had. Thus, Freud's attributions of timelessness to unconscious constellations of memory, phantasy, and wishful motives are inconsistent with his valid claim that these constellations can be modified by psychoanalysis. Psychoanalysis brings the modifications about by facilitating the withdrawal of the libidinal and aggressive energy that invest the provocative repressed wishful motives and allowing the aggression and libido to wind its way forward, as it were, to invest libidinal and aggressive reality-bound projects of adult life or aesthetic experience. The memories in question remain the same, except that the wishes they generate have become quiescent and are now correctly designated in language by the past tense: "I wanted to get rid of my baby brother", instead of "I want to get rid of *x*" where *x* represents displacements of the original object. The irony of Freud's use of the term "timeless" to characterize unconscious contents and processes is that it makes the timelessness and immortality of the psyche directly proportional to the extent to which the psyche is neurotic. No doubt, Freud's atheistic stoicism could have taken an ironical satisfaction in this unstated implication, but Freud never stated this implication or worked out its consequences. Furthermore, his romanticizing narcissism could have caused him to have had in mind the untenable idea of archaic inheritances and the unconscious processes associated with them as emblems, or even evidence, of human immutability. Perhaps it was this narcissism that caused Freud to exaggerate the observational power of clinical psychoanalysis and disregard the findings of evolutionary biology when Jones (1957) asked him to modify his hypothesis of an archaic heritage at the time of Freud's publication of *Moses and Monotheism* (1939a).

The subtle exaggeration in Freud's use of the term, and his preoccupation with Kantian subjective idealist ideas, might also suggest an element of wishful thinking. Freud (1923b) had already identified the superego with Kant's Categorical Imperative. A more correct identification would have been with moral reason and will, since, for Kant (1788), categorical imperatives are the instruments by which moral will acts, just as the superego is the source of the demand that we act according to ideals and moral obligations. Moral will, according to Kant, is also the expression of the noumenal (immortal) in man. The philosophical foundations of Kant's moral theory are completely at odds with Freud's developmental explanation of the origins of conscience and of deontological morality. Was Freud's (1923b) assimilation of the superego to an antithetical philosophy inspired, despite his atheism, in the obscure depths of the unconscious by a narcissistic need seeking satisfaction in the possibility of psychic immortality? It is possible, but given the fundamental premise of psychoanalysis that the brain is the organ of the conscious and unconscious mind, and owes its existence to it, the timelessness and immortality of unconscious processes would have to be bound by the finitude and mortality of the brain that sustains them. It is possible that Freud's narcissism protested against a truth, despite his deep awareness of it, precisely on account of his discovery of psychoanalysis and notwithstanding his knowledge that his discovery had offended human narcissistic hopes no less than those of Copernicus, Galileo, and Darwin.

Freud's theory, with its hypothesis of timeless unconscious elements and processes, can be easily corrected by simply omitting his attributions of timelessness, which his theory does not need, or by limiting the definition of timelessness to the four essential factors identified above: the durability of unconscious organizations, the inability of repression to divest them of motivational efficacy, their indifference to time, and their irremediability by time.

In the end, to the biologically driven narcissistic anxiety about death, there are added the accumulated lost satisfactions of object love. We are driven to seek out "intimations of immortality" not only to ward off our helplessness in the face of inevitability, but also to console ourselves with phantasies of future opportunities to find the satisfactions and happiness that life has denied us on account of failed object relations in which they might have been, but were not, found. Often enough the intimations involve a turning back to childhood, or even before it, as in Plato's (Meno, Phaedo, Phaedrus, Republic) theory of knowledge used by Wordsworth (1805) in his poem "Intimations of Immortality",

> Our birth is but a sleep and a forgetting:
> The Soul that rises with us, our life's Star,
> Hath had elsewhere its setting,
> And cometh from afar:
> . . .
> But trailing clouds of glory do we come
> From God, who is our home. [ll. 59–66)

Compare this poetry of narcissistic and object longing with these lines motivated by temporal object love from a poem by Marvell (1681), "To his Coy Mistress"

> But at my back I always hear
> Time's wingèd chariot hurrying near;
> And yonder all before us lie
> Deserts of vast eternity.
> . . .
> The grave's a fine and private place,
> But none, I think, do there embrace. [pp. 744–745]

Our attitude to time betrays our attitude to death; the degree to which we seek narcissistic consolation in a life after death is inversely proportional to the consolation of object love in life that we have known. Unconscious phantasies may temporarily possess a virtual immortality, but the truth, from which we seek to escape, when the satisfactions of object love are not enough, is that immortality is itself only a phantasy.

References

Aristotle (1947). Metaphysics. In: R. McKeon (Ed.), *Introduction to Aristotle*. New York: Modern Library, 1947.

Arlow, J. (1984). Disturbances of the sense of time--with special reference to the experience of timelessness. *Psychoanalytic Quarterly, 53*: 13–37.

Augustine (c. 397). *The Confessions of St. Augustine*. E. B. Pusey (Trans.). London: J. M. Dent, 1907.

Descartes, R. (1641). *Meditation, Philosophical Works of Descartes, 1* (pp. 133–199), E. S. Haldane & G. R. T. Ross (Trans.). New York: Dover, 1955.

Freud, S. (1896a). Heredity and the aetiology of the neuroses. *S.E., 3*: 191–221. London: Hogarth.

Freud, S. (1900a). *The Interpretation of Dreams. S.E., 4–5*. London: Hogarth.

Freud, S. (1915e). The unconscious. *S.E., 14*: 166–215. London: Hogarth.

Freud, S. (1918b). *From the History of an Infantile Neurosis. S.E., 17*: 7–122. London: Hogarth.

Freud, S. (1920g). *Beyond the Pleasure Principle. S.E., 18*: London: Hogarth.

Freud, S. (1923b). *The Ego and the Id. S.E., 19*: 12–66. London: Hogarth.

Freud, S. (1933a). *New Introductory Lectures on Psycho-analysis. S.E., 22*: 7–182. London: Hogarth.

Freud, S. (1939a). *Moses and Monotheism: Three Essays. S.E., 23*: 7–137. London: Hogarth.

Hanly, C. (1997). On the psychoanalytic idea of time. *Samiksa, 51*: 1–9.

Hawking, S. (1988). *A Brief History of Time*. Toronto: Bantam.

Jones, E. (1957). *The Life and Works of Sigmund Freud*, Vol. 3. New York: Basic Books.

Kant, I. (1781). *Critique of Pure Reason*. N. K. Smith (Trans.). London: Macmillan, 1950.

Kant, I. (1788). *Critique of Practical Reason*. T. K. Abbott (Trans.). London: Longmans, Green, 1948.

Marvell, A. (1681). To his coy mistress. In: H. J. C. Grierson & G. Bullough (Eds.), *The Oxford Book of Seventeenth Century Verse*. Oxford: Clarendon, 1934.

Orgel, S. (1965). On time and timelessness. *Journal of the American Psychoanalytic Association, 13*: 102–121.

Parmenides (1962). The proem. In: G. S. Kirk & J. E. Raven (Eds.), *The Pre-Socratic Philosophers*. Cambridge: Cambridge University Press.

Shakespeare, W. (c. 1595). Sonnet 4. In: *The Complete Works of Shakespeare*. Oxford: Odhams and Basil Blackwell, 1947.

Wordsworth, W. (1805). Intimations of immortality from memories of early childhood. In: S. Gill (Ed.), *Selected Poems*. London: Penguin, 2004.

CHAPTER THREE

Why did Orpheus look back?[1]

Michael Parsons

Winnicott's (1971, p. 38) statement, that the aim of psychoanalytic therapy is to develop a capacity for playing, is famous. His wife commented that Winnicott's work was founded on his own capacity to play, which was part of his way of relating and being related to, and which was there in "his whole style of life". She says "It seems important to note that in his terms the capacity to play is equated with a *quality of living*" (Winnicott, C., 1989, pp. 2–3). She quotes Winnicott's own statement that "Playing is an experience, always a creative experience, and it is an experience in the space–time continuum, a basic form of living" (Winnicott, D., 1971, p. 50). In a lecture in 1963, Winnicott said,

> It is work with borderline patients that has taken me (whether I liked it or not) to the early human condition, and here I mean to the early life of the individual rather than to the mental mechanisms of earliest infancy. [Winnicott, 1965, p. 235]

The idea emerges that central to psychoanalysis is the attempt to help patients develop their capacity for living.

This might not seem exceptional. Who would disagree? In fact, it is radical. Psychoanalysis, having begun as a treatment for neurotic symptoms, came to encompass character disturbance, perverse and borderline conditions, and psychotic states. All these affect the quality of patients' lives. But the assumption may be that analysis improves the quality of patients' living by the way that it deals with such disorders. In this light, change in how patients experience their lives is not incidental in the sense of being unimportant, but it appears as a secondary consequence of the primary work of analysis, which is the diagnosis and treatment of pathology. The radical shift that is implied in Winnicott's work, and which I wish to emphasize and extend, is to take as a central focus of psychoanalysis the way in which people experience their living, and to give primary status to the question of how a person's aliveness is impaired. Neurotic and character disturbances can be viewed not as pathological entities requiring specific treatment, but as manifestations of the way that a person is not managing to live as alive a life as he or she might be able to.

In the chapter "Creativity and its origins", in *Playing and Reality*, Winnicott (1971, pp. 65–85) describes two different ways of living. One is a state of compliance with external reality, "the world and its details being recognised, but only as something to be fitted in with or demanding adaptation". The other is a relationship to reality of "creative apperception". "Apperception" is the perception of something in relation to one's past experience, and, thus, the perception of the inner meaning that something has for oneself. Enid Balint, recognizing that she was talking about the same thing, preferred the phrase "imaginative perception", which she described as "what happens when the patient imagines what he perceives and thus creates his own partly imagined, partly perceived, world" (Balint, 1993, p. 103). Winnicott's purpose in outlining these two ways of living was to propose that psychological health or illness is a function of how far one is living creatively on the one hand, or by compliance on the other. And it does go this way round; that is to say, it is not that our quality of living depends on the level of psychological health we have achieved, but, rather, that our psychological health, or lack of it, is a manifestation of the quality of living we have achieved. In Guntrip's description of his analyses with Fairbairn and Winnicott, he writes that Winnicott once said to him, "We differ from Freud. He was for curing symptoms. We are concerned with living persons, whole living and loving" (Guntrip,

1975, p. 153). I do not think Freud was concerned only with curing symptoms, but Winnicott was making a point about how he viewed the nature of psychoanalysis. There is another chapter in *Playing and Reality*, titled "The place where we live" (Winnicott, 1971, pp. 104–110). For years, I understood that phrase to mean "the area where our experience is located", reading it in the same way as "the house where I live". But I have realized it can also be read as "The place where we LIVE", meaning the place in which, when we manage to be in it, we become fully alive.

"Living" is a "doing" sort of word, and to speak of someone's "quality of living" could seem to refer to how they are getting on with their life, or what they are doing with it. That is important, of course, but what I am specifically focusing on is captured more exactly by the phrase "quality of aliveness". Aliveness is a state of being, out of which the activity of living can grow. As Thomas Ogden writes,

> I believe that every form of psychopathology represents a specific type of limitation of the individual's capacity to be fully alive as a human being. The goal of analysis from this point of view is larger than that of the resolution of unconscious intrapsychic conflict, the diminution of symptomatology, the enhancement of reflective subjectivity and self-understanding, and the increase of sense of personal agency. Although one's sense of being alive is intimately intertwined with each of the above-mentioned capacities, I believe that the experience of aliveness is a quality that is superordinate to these capacities and must be considered as an aspect of the analytic experience *in its own terms*. [Ogden, 1995, p. 696]

Expressions like "the capacity to be fully alive" and "the experience of aliveness" may arouse a sense of recognition in us, but can we be more specific about what they mean, in terms that would let us think how to work towards them?

This brings me to the question of time. I want to put forward the idea that to be fully and creatively alive means living at a point of intersection between time and timelessness. The significance of my title lies in the idea that Orpheus cannot resist looking back at Eurydice because he cannot allow timelessness and time to intersect.

An analyst's first association to the idea of timelessness is likely to be the timelessness of the unconscious. The most direct, immediate contact with the unconscious is in the experience of dreaming. Some patients do not dream, or, more accurately, do not allow themselves to remember dreams. Or dreams might not come to mind during sessions, or there might be dreaming during holiday periods only. In such cases, what is it that is being avoided? To describe a dream belongs to the temporality of daily life, but the experience of dreaming lies outside this ordinary temporality, because dreams are a direct conduit into the unconscious. In dreaming, the time-governed structures of everyday life, by which people live but by which they also defend themselves, are laid open, without possibility of control, to unstructured timelessness. To tell a dream in the analytic situation is more than narration. It is to expose one's dreaming, and this is to place oneself where time and timelessness collide.

Everyday existence within time has a linear structure that gives it, at least in principle, a predictability. It is like those mathematical puzzles that present a sequence of numbers and ask what number would come next. Whether or not you can detect the logic of the sequence, you believe that each number does bear a relation to the numbers before it. Temporal existence allows one to have faith that at least some aspects of one's life are predictable and potentially manageable. In conditions of timelessness, this disappears. There is no before or after. One event cannot be the consequence of another, or have any influence on another. What happens is unpredictable and outside any kind of control. No wonder some patients avoid the intersection of time and timelessness that is involved in the reporting of a dream.

It is difficult sometimes to indicate to patients that they might talk about their dreams without necessarily trying to work out what they mean. There is such a temptation to avoid the unboundaried, open-in-all-directions unpredictability of the unconscious that patients often try to reduce the timelessness of their dream experience to the linearity of an explanation. It might also be difficult for analysts to allow meaning to emerge without constraining it into an interpretation that reduces timelessness to time.

However, dream is not just a message in a bottle or a cryptogram to be deciphered. It is an experience in a person's life. The aim of analysts should be to help that experience expand and enlarge itself in a person's mind in such a way as to enrich their life. This raises important questions of technique, which I have discussed in two recent articles (Parsons, 2006, 2007). My point just now is to say that working with patients' dreams means being interested in, and trying to facilitate, what happens at the crossing-point of time and timelessness.

Bion, in his book *Learning from Experience* (1962), extended the idea of dreaming beyond the sleeping state. He says that a certain kind of internal work, which he calls alpha-function, is required to allow emotional experience to be used for thinking: to produce thoughts, that is, as opposed to mere mental happenings. This applies in the same way both in sleep and in waking life. A person without alpha-function might have psychic experiences while asleep, but these will not amount to dreams. Awake, such a person might hallucinate, or try to evacuate indigestible mental events, but cannot think. Where sleep is concerned, we are used to speaking of dream-thoughts. Bion's point, that alpha-function is needed in the same way in both the waking and the sleeping state, leads to his apparently paradoxical notion that a person should be able to dream while awake just as much as when asleep.

Ogden has made particular use of this idea, clarifying it further in the process.

> For . . . Bion, dreaming, if it is to merit the name, must involve unconscious psychological work achieved through the linking of elements of experience . . . in the creation of dream-thought. This work of making unconscious linkages . . . allows one unconsciously and consciously to think about and make psychological use of experience. [Ogden, 2003, p. 19]

This is in a direct line with Winnicott and Balint. Creative apperception involves the work of forming unconscious links between what is happening now and a person's previous life experience, links that give the present experience texture and complexity that it would otherwise not have. Imaginative perception means linking what the imagination conceives inwardly with what is perceived outwardly, to yield a world that is both perceived and imagined; or, as Winnicott said, both discovered and created. In Bion's terms, this is dreaming, and it goes on not just in the timelessness of night, but whatever might be the time of day.

Telling one's analyst a dream is a bid for aliveness. In response, the analyst needs to stay poised between time and timelessness, and to try to help the patient do the same. Nevertheless, the canvas is much broader than the clinical situation. If the quality of aliveness depends on a capacity for dreaming and reflecting on one's dreaming, awake as well as asleep, this means living one's life at the crossing-point of timelessness and time.

Here is an episode from a clinical seminar. It is what happened in the seminar room, not in the analysis being presented, that I want to highlight. This is an example from life, therefore, not the consulting-room. But, arising as it did in the course of a case discussion, it also reflects the importance of this idea for clinical work.

A candidate presented the case of a man whose childhood and adolescence had been very disturbed. He was abandoned in his earliest years by first one parent, then both. After being looked after by a muddle of grandparents and neighbours, he was reclaimed by one parent and subjected to a bizarre, eccentric lifestyle. He escaped to the other parent, by now a stranger, only to be treated more chaotically still. Particular features of the story were extremely disturbing. Eventually, he moved away and tried to put his life together, but became quite ill. Now he had found his way into analysis. I began the discussion by asking, "What has it been like, listening to this?" The seminar group felt overwhelmed by the absence of structure and boundaries, not just in this man's life, but in their own experience of listening to the candidate's account. The elements of which this man's nightmare was made up seemed to have no relation to each other. It was a random kaleidoscope of horror. The unpredictability had been terrifying for him, and there was also something unnerving for the group about listening with no idea what kind of shock might be coming next.

We tried to relate, as best we could, the listening experience of the group to the details of the patient's story, and the seminar had been going on for about forty-five minutes when one of the members asked, "How old is he?" This was a remarkable moment. A patient's age is basic information, mentioned in the first sentence of a presentation. If it is not, someone will always make a point of asking. Without it, one does not know how to envisage the patient. Yet, we had discussed this history for three quarters of an hour and nobody had mentioned this man's age. We realized that the absence of temporal structure in the patient's experience had conveyed itself into the candidate's presentation and the group's listening, so that we were unconsciously caught up in it as well.

The presenter then answered that the patient was in his late thirties. This was another jolt. We had all assumed that we were hearing about someone at least ten years younger. Questions were immediately asked about the history of these unexpected years. How had he lived? And where? For how long? Sexual relationships? Work? It was as though we were shocked, in the group, by suddenly recognizing how we had fallen out of time and were trying to scramble back into it.

The discussion then became very interesting. The historical structure of the patient's life, indicating what he had and had not managed in terms of psychic development, was obviously important. Our lack of attention to this had been a real failing. But I tried to help the group to avoid now using the patient's history as a refuge from the disconcerting timelessness in which we had been caught. I thought we needed to savour the quality of both at the same . . . time. What mattered was to keep this interplay alive in the discussion. Holding myself, and trying to hold the group, at this point of intersection called for a particular internal poise on my part. I needed, in Bion's terms, to dream the seminar, so as to help the seminar group dream the patient: just as analysts need to dream their patients, so as to help their patients dream themselves. (I am grateful to the candidate presenting the case for permission to mention it and to the members of the seminar for their agreement to my reporting the discussion.)

Some might think it a strange idea that analysts should want to help patients live in a dream. But dreaming, in this sense, does not mean being disconnected from reality. It means being more deeply connected to it by perceiving it through the lens of one's imagination. There is an important link here with reverie. Psychoanalytic reverie, likewise, is not a withdrawn state of semi-conscious vagueness, but a contemplative openness which allows outward and inward reality to interact creatively. There is not time today, however, to do more than notice that reverie also belongs in this conceptual cluster of creative apperception, imaginative perception, and dreaming.

It is because Orpheus cannot dream Eurydice that he has to look back at her. She was his wife, and when she was bitten by a snake and died, he went down into the underworld to reclaim her. Hades, king of the underworld, was captivated by Orpheus's singing and agreed that he should lead Eurydice back into the world of the living, on condition that he did not turn round to see her on the way. Orpheus, though, could not resist looking behind him to check that she really was there. What he saw, because he had looked back, was Eurydice helplessly disappearing again forever.

The darkness of the underworld, home to the disembodied spirits of the dead, where no events unfold, can represent for us the timeless unconscious into which our dreams at night drop a plumb-line. The world above is structured in time. There lives are lived and one thing does lead to another. The two worlds appear utterly separate, and the desire of Orpheus to recover his wife from the one and get her back into the other emphasizes the disjunction between them. The myth of Proserpine, who had to live half the year in Hades and the other half on earth, recognizes that the underworld cannot be obliterated; but the alternation of time and timelessness still leaves them disconnected from each other. Hades' instruction to Orpheus points in a different direction. He seems to say, "You can take Eurydice back with you to the world of light and life, but not if that means leaving this world behind. If you have to look back with your material vision to check, in concrete terms, that Eurydice is there, she will not be there. Create her with the inner vision of your imagination, and she will be there externally as well."

Eurydice represents Orpheus's aliveness. Not being able to allow timelessness and time to intersect, he could not dream her. The return of Eurydice to the underworld symbolizes the loss of imaginative and creative aliveness that results from not being able to stay poised at that point of intersection.

Note

1. This paper is revised from one given at the EPF Conference in Barcelona on 1 April 2007.

References

Balint, E. (1993). *Before I was I: Psychoanalysis and the Imagination. Collected Papers of Enid Balint*, J. Mitchell & M. Parsons (Eds.). Free Association Books and Guilford Press.

Bion, W. (1962). *Learning from Experience*. London: Heinemann [reprinted in *Seven Servants*. New York: Aronson, 1977].

Guntrip, H. (1975). My experiences of analysis with Fairbairn and Winnicott. *International Review of Psycho-Analysis*, 2: 145–156.

Ogden, T. (1995). Analysing forms of aliveness and deadness of the transference-countertransference. *International Journal of Psychoanalysis*, *76*: 695–709.

Ogden, T. (2003). On not being able to dream. *International Journal of Psychoanalysis*, *84*: 17–30.

Parsons, M. (2006). The analyst's countertransference to the psychoanalytic process. *International Journal of Psychoanalysis*, *87*: 1183–1198.

Parsons, M. (2007). Raiding the inarticulate: the internal analytic setting and listening beyond countertransference. *International Journal of Psychoanalysis*, *88*: 1441–1456.

Winnicott, C. (1989). D. W. W.: a reflection. In: *Psychoanalytic Explorations*. London: Karnac.

Winnicott, D. W. (1965). *The Maturational Processes and the Facilitating Environment: Studies in the Theory of Emotional Development*. London: Hogarth.

Winnicott, D. W. (1971). *Playing and Reality*. London: Tavistock.

CHAPTER FOUR

Unconscious memory from a twin perspective: subjective time and the mental sphere

Jean-Claude Rolland

Whenever we set up a psychoanalytic situation, what else are we doing but having, first of all, recourse to the formal notion of time as measured by clocks and calendars? We say to the patient, "We'll be meeting on such and such a day, at *x* o'clock, and each session will last *y* minutes; there won't be any sessions during my holidays—I'll tell you the dates well in advance—or on public holidays." We add that he or she will have to pay for all sessions when the analyst is available; this, therefore, excludes any subjective freedom, flexibility, or arbitrariness with regard to this physical time that we make into an implacable reality.

However, once that temporal setting is implicitly established, with its aim of conferring optimal effectiveness on the analytical method, there begins to emerge a vast number of temporalities that have nothing to do with real time; they blur our traditional experience of time and might even appear to "suspend its flight", replacing it not only with the obsolescence of anachronism, but also with a mysterious atemporality, given the evanescence of every means and of every desire to measure it. When, with quite surprising consistency, analysands about to embark on psychoanalytic treatment ask, "How long will the analysis last?", the only answer I can give—because it seems to me to express in the simplest and most accurate terms what I have to say on the topic—is, "Well, you'll see, once we begin, that will no longer be an issue."

And it is true to say that the question of the length of treatment does lose all interest both for the analysand and for the analyst. That does not mean that they—and in particular the analyst—are unaware of the fact that the analysis will come to an end one day; accepting the idea that it might be interrupted from one day to the next is an essential condition for going on with it. The lack of interest in the possible length of an analysis comes from the fact that its unfolding depends less on the logic of time than on that of space: the mental space through which free associations and interpretations have to travel in all directions in order to compensate for the fault-lines that separate the psychic agencies, to lift libidinal repression and to undo fixations to Oedipal objects. That is why, in "Analysis terminable and interminable", Freud (1937c) laid more emphasis on the complexes—the "underlying bedrock" of femininity and castration—than on the actual length of treatment when he discussed the factors that could hinder its accomplishment. It is, of course, true that every undertaking requires temporality, that nothing can exist timelessly: only "nothing" is unaffected by time, the conclusion that *in fine* I shall come to. None the less, temporality is no more than an incidental and contingent feature of mental phenomena; it is not the cornerstone of the fulfilment of wishes or fantasies, for example, or that of dream-work, or of what takes place during an analysis. The distinction is a slight, almost artificial, one, but if our approach to this question of time is to remain analytical, we have to keep it in mind with all the obstinacy that we can muster.

Perhaps, in some distant future, English may become the official language of psychoanalysis, yet no analyst can think about and practise psychoanalysis other than through his or her native language, because it is this that gives analysts their world vision and a certain kind of organization in connection with consciousness and the unconscious in their mental life. Language is the instrument both of repression and of expression: of what can be made conscious. No other category of thought is subject to as many variations depending on the idioms used. That argument was put forward by Emile Benveniste, whose work is of some considerable significance for psychoanalysts, in his discussion of "linguistic time". According to Benveniste (1966),

> It is a mistake to think that the temporal system of any given language mirrors the nature of "objective" time; that mistake springs from the propensity to see in language an exact image of reality. In fact, however, languages offer us only a variety of constructions of reality, and it is perhaps precisely in the way in which they construct a complex temporal system that they diverge the most from one another. [p. 5]

To focus strictly on the semantic level, three examples seem to me to illustrate this problem, which represents a real difficulty in psychoanalytic communication.

The title of Freud's paper "Die endliche und die unendliche psychoanalyse" ("Analysis terminable and interminable") is translated literally into French as "Analysis with end and analysis without end". The word "end" (*fin*, in French) has two meanings that inevitably resonate with each other whenever the word is used. It designates not only the cessation of an act, its conclusion, but also its finality, explicit or secret. In addition, the French word *fin* is homophonous with *faim*, which means "hunger" or "appetite" (for life, for learning, for changing), and in everyday use this both widens and obfuscates its linguistic domain. It is, thus, impossible for French readers not to confer on this highly important paper of Freud's a significance linked to the very ambiguity it contains; they cannot avoid wondering about Freud's true intention in writing it. Was he exploring the ending of an analysis, from a procedural and technical point of view? Was he meditating, somewhat disappointedly, on what the analytical method can and cannot accomplish? Was it a tortuous way of exploring the libidinal energy that the analysand puts into the treatment, on which depend the quality of the transference, the wish for analysis, and, therefore, the amount of time that he or she is willing to devote to it?

My second example concerns the word *Erinnerung*, which is translated as both "memory" and "remembering" (or "recollecting"). The etymological peculiarity of the word, which derives from *innern*, meaning "inside", has often been pointed out. Hence, if we were to take it literally, *Erinnerung* designates the act of putting "inside" something that, by implication, is "outside". This idea of internalization, of the work of internalization, that describes what memory does, is not immediately obvious in the corresponding words in French, and I do envy the ingenuity of the German language that is so lacking in my mother tongue. It is possible that the fact that Freud did have available to him that kind of linguistic instrument enabled him very quickly to understand that the preservation of a traumatic childhood memory which alienates the adult's mental development does not, in essence, have to do with *time that has stopped*, but with *a fragmented mental space*, parts of which define the ego while others compel us to acknowledge a space that lies outside the ego. The contrast "time / timeless" becomes clearer only when this aspect "ego / outside the ego" is taken into consideration, in the sense that the former has to do exclusively with present time while the latter involves, just as exclusively, an atemporal dimension. That contrast has as its common umbilicus a third dimension, the timelessness of unconscious fantasy, which Freud called *Zeitlos*.

My third example is that of *Nachträglichkeit*, "après-coup" ("deferred action"), which I would link to "screen memory", *Deckerinnerung*. In its ordinary, present-day use, according to one of my German colleagues, *Nachträglichkeit* implies going insidiously back over a reproach addressed to someone in order to take one's revenge, and has connotations of bearing a grudge, of resentfulness. The vocabulary Freud used certainly did not come from a dead language, and it is well known that he preferred words in everyday use to scholarly designations.

One thing that does strike us is the fact that, whenever patients devote their free associations to the work of remembering in order to recollect those elements that go all the way back to their childhood, they do not make a direct narrative of these events, but substitute for them more recent memories—even present-day ones—which represent and anticipate them. Their memories work along substitutive lines, following an analogical method, by recognizing in recent events earlier elements of which they were not aware. A man will complain, for example, about the hurt he felt when his wife made some unfortunate remark, and, under the effect of the transference, will go on to think of the time when his mother refused to lift him up into her arms. Only the analyst's interpretation will succeed in restoring the full chronological truth to that immemorial emotional reality.

In psychoanalysis, then, memory revives in the present, retroactively, as it were, events that, although they took place in the past, do not belong exclusively to the past; in so doing, it appends new layers to pre-existing memory traces. Thereafter, an interpretation can methodically sift through this neo-structure and uncover the traumatic memory. It is surprising to observe the extent to which memory disorganizes time, mixing up the categories of past, present, and future, of *nach* and *vor*, for its own use, just as it abolishes, indeed, from a spatial perspective, the distinction between image and screen. It is undoubtedly true to say that Freud's concepts of "deferred action" ("après-coup") and "screen memory" are one way of defining—a roundabout way, perhaps, but, in the end, highly precise—the temporal flux of mental life much better than a direct reference to time, *Zeit*, could ever do.

Indeed, it is always in an incidental manner and never without some hesitation that Freud confronted issues concerning temporality. He did so mainly in his metapsychological papers, first in "The unconscious" (1915e), then in *Beyond the Pleasure Principle* (1920g). The conclusion that he reached, however, was quite categorical. This is what he says:

> The processes of the system *Ucs.* are timeless; i.e. they are not ordered temporally, are not altered by the passage of time; they have no reference to time at all. The reference to time is bound up, once again, with the work of the system *Cs.* [1915e, p. 187]

That assertion brooks no contradiction. Nevertheless, with the deeper understanding we have nowadays of the psychoanalytic experience, we are perfectly entitled to ask whether it ought to be somewhat nuanced, particularly with regard to the relationship between time and "the work of the system *Cs.*", in so far as we now have a better idea than Freud did of its complex and heterogeneous nature.

One of my patients, after some considerable time in analysis, had the following dream, which terrified him: the whole length of his arm was ripped open, the cut going deep down. His muscles were laid bare and—an even more unpleasant detail—you could see the pale whiteness of the aponeurosis. A feeling of paralysis weighed heavily on his attempts at free association, yet, towards the end of the session, he was able to recall the phlebotomies to which his mother repetitively subjected herself, terrifying his father and his sister and exacerbating the fascination which his incomprehensible certainty that his mother was irredeemably unhappy had on him as a three-year-old boy. That was the first time that such a recollection had emerged, and it left the patient speechless for some time. I felt that the stupefaction that struck his power of speech in the session was simply the return of the bafflement that had overwhelmed him as a young boy, but this had immediately been struck by amnesia, so that no subjective experience of it had been possible. Its revival as an after-effect in the transference both reproduced that anachronistic moment of his childhood and, since it was put into words in the here and now of the patient's discourse, turned it into a past event.

When he was once again able to talk, the patient discovered something else. Some years later, when he was ten, his mother, henceforth less overwhelmed and somewhat ashamed of her scars, would hide them by wearing very showy bracelets. The patient recognized in the colour of the aponeurosis that he saw in his dream that same pale whiteness so typical of her scars. Suddenly, he recalled the attitude that he had adopted towards these: he knew of them, he knew they existed, but he did not "see" them. In exactly the same manner, when his mother died of a brain haemorrhage a short time before that session, he did not "see" in the X-rays which his mother's doctor showed him (the patient is also a medical practitioner) the traces of the lesion. In that way, the denial of perception, so typical of his childhood and the passions of the Oedipal phase, was able, in the adult he had by then become, to be revived in the present; thus, in a way similar to *fueros* in a Spanish court of law, it could be incorporated into the physical time that the reality principle imposes on mental life. Within the present time of the patient's discourse, the enunciation enabled this past present to be converted into a past historic.

The patient arrived late for his following session, the next day. He was in a bit of a hurry, and told me that he had put himself to a great deal of trouble in order to reach my consulting-room. In a somewhat feverish tone, he reported the dream he had had that night. In it,

> he was landing on one of the Normandy beaches with the Allied troops. He was at the front of the boat, which opened up to facilitate

> the manoeuvre. He was on his feet, the landing was no easy matter, and he felt afraid.

He associated the dream to a film he had seen shortly before. When I commented that he probably had the same kind of feeling when he "put himself to a great deal of trouble" getting here, the patient replied: "Exactly", immediately adding that, although he did not know why, that dream seemed to have something to do with the one he had reported during the previous session. I was struck by the fact that the temporality at work in this dream—doubly anachronistic in that it evoked a time in history when the patient was not even born and that its underlying motive was clearly playful and infantile—was reproduced as such in the time of the transference, and that my interpretation opened up for him unexpected depths. Then, once more, the flow of free associations dried up, and the dream seemed to fade out of sight. We came back to it, however, in a roundabout way. Some other closely relevant material brought to mind the island from which his family originated and to which—because he was in conflict with his father—the patient had no intention of returning. He did, however, recall the holidays he had spent there with his parents on a regular basis: the departure, the ferry, and above all the arrival, which was a very emotional moment for him; the passengers had to wait until the hull of the boat opened to let the vehicles out. The patient could not remember what that manoeuvre was called. I said that the word he was looking for was "disembarkation" or "landing", adding that the dream he reported probably had something to do with that event in his childhood. That interpretation astonished him and led him to recall an incident which, he said, had never come to mind in spite of the tremendous impact it had had on him: during one of these disembarkations, a boy of about his own age, a close friend of his who was standing next to him on the boat, had had his arm seriously wounded. The whole length of the boy's arm was ripped open, and the muscles and strips of skin were visible, exactly the image he had seen in his previous dream.

In this way, specifically anachronistic temporalities control different memory complexes that not only share a similar sexual and traumatic representational content, but also follow on from one another according to a chronology that is seemingly whimsical in so far as it is independent of the historical circumstances in which they were produced. This sequence must, therefore, be determined by something other than temporality. The system *Cs.*, to borrow Freud's term, does not function in the same way in all of the mental *loci* in which these different memory complexes are to be found. As this observation makes clear, these complexes are not repressed; they have access—with more or less difficulty and by following somewhat roundabout paths—to remembering and consciousness. In addition, as they progress through mental space, they leave behind their propensity to be conjugated in the present tense and take the form of a past tense of time that follows the precepts of the reality principle.

They belong to the preconscious (*Pcs.*) system. When we think of the development of psychoanalytic thinking, we really ought to pay more attention to the history of the concept of the preconscious. Freud himself more or less left it to one side as soon as he was able to replace his descriptive model with a structural one. He then identified it with the ego and the *Cs.* system, on the perfectly justifiable basis that, in each element that goes to make it up, thing-presentations are closely linked to word-presentations. However, if we stay with the topographical point of view, which is very much to the fore in the transference experience of an analysis, we cannot but acknowledge the fact that the space of the preconscious is infinitely greater than that of the ego. From the ego at the centre to its most outlying frontiers, many variations in the subjective feeling of belonging can be observed, as well as in the capacity of the instances that compose it to become conscious. The word "off-centre" would perhaps be an appropriate description of the geography of the preconscious domain. In the mass of memory experiences that make it up, some can be directly recalled; these can thus immediately take their place in the tenses of language: "I felt loved by my mother until my brother was born". Others, however, demand a slow, lengthy process of recollection and bear the hallmark of the time in which they occurred historically, remaining locked in the moment when they emerged, far removed from the official ego. In order to become part of the movement of subjectivity, they require this "time of the event" to be transcribed as "chronological time". Dreams and their interpretation are the fundamental agents of that mediation. In the "landing" dream that I quoted earlier, a memory trace was brought into present time, formulated as "I am disembarking". The interpretation that followed on from it restored its temporal dimension and turned it into a memory. It could then be formulated in terms of "When I disembarked with my parents . . ."

This activity of recollection is the task—not in itself sufficient but none the less necessary—of what we call the work of the analysis. It is not an end in itself, as was for Marcel Proust the aesthetic aim he had in mind when he wrote *Remembrance of Things Past* (*In Search of Lost Time*). It consists in opening up for free associations a path that will lead to the frontier that separates the preconscious from the unconscious. At that point, a further task awaits, more arduous and very different in the way it progresses: the lifting of repression, the suspension of that element which is the opposite of memory, infantile amnesia. It would not be sensible—indeed it would hardly be possible—to dispense with this "preliminary work", the *Vorarbeit* of every analysis. The need for this to be accomplished derives from the very laws that organize preconscious memory. This is not a matter of a passive and mechanical depositing of the perceptual and emotional experiences that individuals have in their relationships with their Oedipal objects. Putting into the individual's memory the historicity of the events that he or she has experienced conforms also to an internal necessity: that of building up, as against the negativity of the unconscious, a stock of representations, a figurative material that aims both to satisfy, through substitution, the individual's insatiable drive-related demands and to give to the ego the means whereby it can free itself from the influence of the unconscious and turn towards the outside world. In that sense, every memory is a *Deckerinnerung*, a screen memory.

Compressing the argument in the manner that I have adopted is, all the same, one way of doing justice to Freud's untiring effort—from his "Project for a scientific psychology" all the way through to *Beyond the Pleasure Principle*—to establish a theory of memory that took into account its twofold subservience: to the reality dimension of which it is a trace, and to the unconscious of which it is the counter-cathexis. Thus, it becomes easier to understand that the *locus* of the preconscious which is devoted to the various components of memory is less subservient to temporal than to drive-related considerations. That is also why the memory trace which is furthest away from—or lies deepest down in—the ego and close to the frontier separating it from the unconscious is not the earliest or the most traumatic of memories; it is the one that has proved to be the best at providing unconscious fantasy with a replacement representation and, consequently, at carrying out its repression.

The fact that preconscious formations participate in the way in which the primary process functions compels them to uphold many different and conflicting interests that considerably loosen their dependent relationship with respect to the ego. This could be compared to the situation of Somali tribesmen who, plagued by religious beliefs and under the covert influence of foreign powers, pay only lukewarm allegiance to central government. Similarly, in so far as we consider the link between a thing and a given word to be the primary "linguistic act", the languages in which these formations are expressed are foreign to each other. In the preconscious, we find all the states of language that a given individual goes through in the slow process of acquiring that of his or her community. This may give us a better understanding of the concept of "translation" that Jean Laplanche suggests for describing those processes characteristic of remembering and the lifting of repression.

Similarly, the time dimensions that we find are plural in number. The preconscious is also a place for preserving the historical periods that have influenced an individual's destiny. We could draw a parallel here with a specific aspect of institutional religion. Mount Athos is the holy place of the Orthodox Church. Every country that belongs to that religious tradition has built a monastery there, one in which, over and beyond the forms of liturgy accepted by all, the traditions and rituals of the originating country are maintained, including the measurement of time as it then existed in that country when the monastery was founded. Anyone visiting these monasteries will be surprised to find that when it is midday in the Russian monastery, it is six o'clock in the morning in the Polish one and six in the evening in the Bulgarian one. The monks in the Russian monastery are hard at work, those in the Polish one are at morning prayers, while the Bulgarian monks are beginning the evening service.

This cosmopolitan nature of the preconscious is not without its charms; it is even somewhat poetic. Its very richness ensures that the individual will feel firmly rooted in the reality of history and the passing of generations. The aim of analysis is not to compel the patient to blend into the temporal dimension of shared reality; it dries up only the part that has to be so dealt with in order to treat the unconscious and its pathogenic offshoots, and that process in itself is often painful and marked by regret. We would like to live forever what we have only briefly experienced: the final impulse of our insatiable longing for eternity.

I am well aware of the fact that teleological reasoning has its limits. That said, I would argue that this multiplicity of temporal modes working together in the preconscious, this "fragmented time" which, according to Green (2000), defines its very nature, provides the psychical apparatus with the material that it needs to overcome the obstacles which the reality of the unconscious places before it in order to hinder its expansion. Obstacle or dilemma: I do not know which word best describes the operation by which a (mental) substance, the unconscious, which Freud defined in temporal terms by its absolute negativity-*Zeitlos*—enters into temporality. That dilemma could not be perceived before psychoanalysis set out to revolutionize our thinking; it lies at the heart of that *praxis*. The lifting of repression, which no other method can accomplish, consists in converting, albeit by roundabout means and leaving some aspects untouched, a formation defined by its negativity into something positive.

In my view, the idea of negativity, which Green (1993) has gone into at some length, enhances that of repression. Unconscious fantasy, like everything else that has its roots in the unconscious, lies not only outside of time, it is non-existent, it has no being, and not only with respect to consciousness. Its non-existence or evanescence has the same relationship to the preconscious as a shadow has to the body, as death has to life. This opens up the question of the co-substantiality of being and time, one which, being more philosophical than psychoanalytic, I cannot go into in any detail. In order not to give in to laziness—or to fear—and abandon the arduous exploration of these unfathomable depths of the human mind, I shall present three illustrative arguments.

The first of these is etymological. The use of the word *rien* ("nothing") in French dates back to the eleventh century. It was built on the (Latin) feminine noun *res*, "thing", and took on a pronominal and semi-negative sense because it was frequently associated with the negative particle *ne* (Picoche, 1971). *Rien*, therefore, designates a thing not *in absentia*, but in its negativeness. So, after all, that really is . . . something! In other words, there is something underneath the nothing that calls for its resurrection. The blurring which covers that natural etymological construction no doubt reflects the obfuscation of our thinking when faced with this category of the negative.

My second example is the somewhat cryptic remark Freud made in a letter to Lou Andreas-Salomé in 1915:

> The unity of this world seems to me something self-understood, something unworthy of emphasis. What interests me is the separation and breaking up into its component parts what would otherwise flow together into a primeval pulp. Even the assurance most clearly expressed in Grabbe's Hannibal that "We will not fall out of this world" doesn't seem a sufficient substitute for the surrender of the boundaries of the ego, which can be painful enough. [Freud, 1915, p. 309]

What indeed is the nature, beyond the frontiers of the ego, of this world of the unconscious into which repression causes the individual to fall?

My third illustration concerns a very well-known passage taken from "A child is being beaten" (Freud, 1919e). It could be said that, with extraordinary acuteness, Freud, in that paper, goes to the very heart of the structure of unconscious fantasy. Here, to borrow Freud's own words already quoted, he separates and breaks up "into its component parts what would otherwise flow together into a primeval pulp". The temporal development of this kind of fantasy is in three stages, two of which belong to the preconscious. The fantasy is there formulated as "My father is beating the child" [*ibid.*, p. 185] in the first phase, while in the final phase the representation of the Oedipal object disappears, leaving the fantasy as "A child is being beaten". Both of these were—or are—accessible to verbalization and to consciousness. In the second phase, however, since the fantasy has succumbed to repression, its nature undergoes a radical change. "The fantasy is accompanied by a high degree of pleasure, and has now acquired a significant content [which is unmistakably Oedipal in nature]. Now, therefore, the wording runs: '*I am being beaten by my father.*' It is of an unmistakably masochistic character" (*ibid.*).

Freud goes on to say something that, for the argument I am developing here, is of the highest importance.

> This second phase is the most important and the most momentous of all. But we may say of it in a certain sense that *it has never had a real existence. It is never remembered,* it has never succeeded in becoming conscious. *It is a construction of analysis,* but it is no less *a necessity* on that account. [*ibid.*, my italics]

Freud did not go any further into the negative aspect of the unconscious that he adumbrated in that extract, particularly when he stated that this second phase of the fantasy "has never had a real existence". It could be said that he came back to this point much later, in his discussion of the death drive and its tendency to unbinding and wiping out psychical formations. This, however, is not quite the same thing. Nevertheless, the fact that, at this point in his work on psychoanalytic theory, he contrasted the idea of negativity with that of construction appears to me to be a particularly interesting road to follow for the practice of psychoanalysis. Unconscious fantasy, which is not adequately defined by the category "outside of" (outside of time, of place), can have access to the positive dimension of time, of place—and therefore of being—only if it links up with a preconscious memory; that is, through being represented by the substance of memory. A construction is the part of our interpretative activity that identifies in preconscious remembering the interplay between a fantasy and its processing. In the clinical illustration I reported earlier, the memory that emerged of the mother's phlebotomies and of the denial that the young boy set up with regard to acknowledging them carries within its many folds a considerable number of fantasy representations that have to do both with Oedipal attachment to the mother and with castration anxiety. It follows that, in analysis, remembering is much more than an activity carried out by memory; it includes a process of renunciation of drive satisfaction.

As Freud pointed out, unconscious fantasy is never remembered. It may, however, be the object of what, with a slight play on words, I would call memorization. By its negativity, the unconscious activates from an unexpected *locus* the work of memory, as if we recall something as much to discover what we once were as to find out what we have not yet become.

The relationship between psychoanalytic thinking and the concept of memory is one of elective affinity. It is true to say that psychoanalytic theory was built on hysteria, but, to be even more precise, we should add that it was on the basis of the highly specific memory problems associated with that disorder that Freud was able to construct his initial ideas on repression, the unconscious, and the pre-eminent role of infantile sexuality in the workings of the mind. During his correspondence with Fliess, he discovered that hysterics suffer from reminiscences (i.e., from an unconscious memory that, avoiding both remembering and forgetting, becomes a reality through the presenting symptoms: a memory that forgets to remember). That discovery was to prove equally valid with regard to the other types of neurosis and psychosis, thus reinforcing Freud's idea that every mental disorder has to do with a problem of memory, or, at least, with a tendentious use of that particular feature of the human mind.

Freud realized that hysterical paralysis is different from its neurological counterpart in that it makes use of an imaginary anatomy; he discovered also that the anamnesis of their illness that these patients give spontaneously refers not to objective historical reality, but to a construct of memories organized in terms of a purely erotic and passionate logic. The only history that neurotic disorders take as their basis is that of love, its desires, its expectations, and its sorrows; the neurotic disorder is a substitute for, or a representative of, that impulse. That intuition was indeed precious for the subsequent development of psychoanalytic theory: the close connection between disturbances of consciousness which affect the representation that hysterical patients have of their *bodily space* and those which influence their *subjective temporality*. In human beings, memory involves both time—that was well known before psychoanalysis came on the scene—and mental space; the latter was a new discovery.

Memory is not simply the set of traces that something belonging to the past has left behind; it tends also, for everything that touches on desire, to be the place where highly crucial events experienced by the individual in the course of his or her development are kept. In order to bring about a reversal of its natural functioning, memory makes use of space. Repression creates a split in the mental apparatus. On the one hand, there are the conscious agencies, such as the ego, in which memory complies with remembering, forgetting, and recollecting; the only temporal dimension it knows is that of passing time. On the other hand, there is the unconscious, which endeavours to preserve infantile passions, with their objects, circumstances, and even disappointments; the only temporal dimension that it takes into consideration is that of time which does not pass, as Pontalis (1997) put it.

For psychoanalytic thinking, the work of remembering is not simply going back in time, it is also a progression through mental space and a forcing of the internal frontiers that provide reminiscences with a sanctuary against forgetting and the pain which that involves. Every aspect of psychoanalytic treatment contributes to that project, for which remembering is the very essence of the work of an analysis: transference, which brings to the fore repressed objects that have been preserved, displacing them on to the analyst; regression, which brings what the analysand says into a system of free associations, liberates it from small talk and activates the capacity to link up with unconscious representations; and even dreams, which are—to use the term that Freud himself used in *The Interpretation of Dreams* (1900a)—*Erinnerungen*, recollections. For a psychoanalyst, talking about analysis implies talking about memory, and vice versa. Let us then go straight to the heart of that experience, into the sanctuary of memories that is the analytical situation.

This patient, who was in his forties, began analysis for a very moving reason: he wanted to free himself from the uncontrollable violence that took hold of him with his partners—both in his personal relationships and in business—whenever the relationship he formed with them became intense, intimate, and lasting. Jealousy and querulousness were sparked off by any perceived event, no matter how trivial, and from then on became explosive, to such an extent that the patient represented a real threat to the other person involved. He realized that he had to do "something" about it if he were not to spend the rest of his life in emotional isolation and social exclusion. A recent incident had really frightened him and made him decide to seek help. That his mental life was organized in a paranoid manner was perfectly clear; it was of the kind that Freud described as neurotic (Freud, 1922b). Shortly after beginning the analysis, I came to the conclusion that some real analytic work would indeed be possible with that patient.

There was, all the same, one somewhat alarming symptom that weighed heavily on the beginning of the treatment: the patient was convinced that one of his uncles, Uncle D, had, during the patient's childhood, sexually abused his younger sister and perhaps also his even younger brother. The patient believed that his family were aware of this "crime", but did nothing about it; he therefore broke off all contact with them, and sent to his family a series of insulting and threatening letters in which he said he would expose that misdeed. The patient had some inkling that this compulsive activity of his was somewhat dishonourable and perhaps even cruelly unjust. That symptom grew in importance all through the early months of the analysis then, without our getting any closer to its unconscious motivations, gradually faded.

The session I am about to report put a stop, for all time, to that *neurotica*. I shall describe what took place as exactly as possible.

> That day the patient's demeanour was less edgy, his voice less tense and he spoke in a more relaxed rhythm. He began by talking not about the plots that he suspected were being hatched against him in his social life, but about what was going on inside him. There was something more human and less petrified about him that day.
>
> After he had spoken about his sadness, loneliness and despair, the patient remembered the dream he had had the previous night. In it, he was talking to his mother. He was telling her about what Uncle D had done, i.e. that he had sexually abused M, the patient's sister, and had also committed murder.
>
> The patient then seemed to pay no further heed to the dream. He spoke about my holiday break, the dates of which I had given him at the end of the previous session, and regretted that it would be such a long one. Then he thought about M, his sister, saying that all she lived for was her work; every one of her love affairs had turned out to be unhappy. He then thought of the most recent relationship that he himself had been in, which had ended tragically through his own fault. I commented that he was thinking about himself and his own unhappy affairs while thinking about M in the dream.
>
> When M and he were children, he added, they were looked upon as the two "big" children as opposed to the two "little" ones, in much the same way as "bad" children are contrasted with "good" ones. He then thought of his mother, saying that he would like to spend a few days'

holiday with her, just to talk things over. I commented that the conversation with his mother in the dream probably had something to do with that, adding that his wish to spend some holiday time at his mother's no doubt had something to do with my own holidays. I went on to say that the "uncle" in the dream was probably a reference to me. I said all of this because the transference activates the "remembering" function of dreams and because, behind each of the images therein represented, there lies a figure from the repressed past.

That series of comments on my part, linking elements common to his free associations and the language of his dreams, took him aback. He said that he had indeed thought about Uncle D recently, in a kind of day-dream: he had imagined that he was asking his mother if he could come to visit her on such-and-such a day; she replied that that was precisely the time when Uncle D would be staying at her place. The patient then imagined himself saying to her, "It's either him or me". I commented that the murder scene in the dream had probably something to do with that day-dream.

Again he was dumbfounded. Then he remembered a scene from his childhood involving that uncle. Uncle D had taken him on holiday to the seaside. One day, when the patient was fighting with his younger cousin, Uncle D's son, and was taking unfair advantage of his strength, Uncle D had thrust the patient's head into the water so violently and had held it like that for such a long time that he really thought he was going to die. The patient then went on to say that he wondered why Uncle D had not chosen some other form of punishment. I commented that the fact of my going away on holiday seemed to him to be like a punishment.

Again the patient was flabbergasted. Yes, no . . . well, anyway, he could appreciate the fact that I did need a break . . . Then there came into his mind an idea that he often had during his sessions without really thinking of putting it into words. He often imagined himself lying on the couch like a recumbent statue, his hands crossed over his heart, as he had seen his two grandfathers lying after their death. I commented that when he thought of his uncle almost killing him, that probably had something to do with me here in the analysis. The patient then spoke of his fear of death, and of his constantly seeking a way out. I commented that perhaps going to his mother's was a way out with regard to my forthcoming absence. He spoke then of his suicidal behaviour, which was by then less violent but could be seen in his tendency to eat and drink too much; he added that he would like his mother, who was interested in dietetics, to help him overcome that.

> Remembering came in waves, with the dream-work helping it along. Since it gave rise to a narrative, the dream brought into the field of verbalization—and therefore into the sphere of the ego—a memory trace that until that point had been located in the space of the unconscious, of the id. A topographical displacement is therefore the first step in the movement by which a memory comes to consciousness.
>
> The signifier "murder" mentioned in the dream represented the punishment inflicted by the uncle which had been repressed. One can only admire the skilfulness and elegance with which the patient's free associations translated it into "it's either him or me". Additionally, that unconscious memory trace was linked to the incestuous fantasy that had long before reached the patient's consciousness and that no doubt represented it there. It was connected also to several unformulated associations that were able to come to light in the transference in the form of a scenario that attributed to me the role of murderer, and to the patient himself that of victim, in a kind of silent memory.

There was something eerily fantastic about that session. The analyst was the ghost of the "murderous" uncle and the analysand, in recollecting the damage inflicted on him, was a recumbent statue, "dead". The masochistic potential that from the outset had erotically saturated the actual punishment—a historically datable event—turned it into a fantasy of the "child is being beaten" type; this lay at the core of the individual myth of the incestuous uncle and facilitated the development of a primitive and paranoid turn of mind.

The work of remembering that free associations accomplish, supplemented by the work of interpretation, activates the displacements, distortions, and transformations that the primary trauma, the historical fact, has to go through in order to reach consciousness and thereafter be cleared away.

Not only is a given historical event registered simultaneously in several *loci* of the mind, so that remembering is diffracted through several locations at the same time, but also several such events can be registered in a single *locus* through the process of condensation, which lies at the heart of the production, in free associations, of similarities and analogies. Remembering thus causes an event to emerge, then the antecedents of that event, and, in the course of analysis, calls for it to be dissected on several levels.

Let us stay for a moment with the patient I have just described and see what occurred some months after that moment of remembering. Towards the end of the session preceding the one I shall now discuss in some detail, the patient discovered that with his maternal grandmother, with whom he had lived when he was between seven and fourteen years of age while his parents were abroad with their three other children, an intensely powerful attraction had formed. The patient then had to come to terms with the fact that he might have been "seduced" by his grandmother: he slept every night in her bed, while his grandfather slept in a small bed in the dingy bedroom. Above all (and this idea frightened him even more), he himself might have "seduced" his grandmother. That idea sprang from the fact that she made no attempt to control him or fix any limits, she just let him live his undisciplined life as a pre-delinquent adolescent.

At that point the analysis had been going on for four years, and the work of remembering was at its acme. When any event occurring at that time, no matter how innocuous, gave rise to surprise or questioning, it was talked about in the sessions, and, through the interplay of free associations and interpretations, enabled new fragments of his infantile memory to emerge. On one occasion, while he was talking feverishly about his relationship with his grandmother, seeming to relive it emotionally (or perhaps even experience it *in vivo*, as though it belonged to the immediate present, without any fading away or historical contextualization), I had to tell him that our time was up and that we would have to stop the session at that point.

He began the following session thus: he had been shocked by my reaction, cutting him off like that in the middle of what he was saying. At the same time, he was surprised by the fact that he felt shocked, because he realized how important and helpful maintaining the setting was for him. That was why he had made no difficulty about stopping. There was another thing, too, that had shocked him: as he left the consulting-room, instead of leaving the front door open as he always did so that the patient after him could make her way in, he had shut it quite abruptly. He immediately realized what he had done, and almost rang the bell to let me know. At that point, he said to himself that he had better write it down so as not to risk forgetting the incident.

For some time previously, he had been strangely mindful of the idea of courtesy between patients. He was proud of the courtesy he usually showed towards the patient who followed him on that particular day of the week; he liked leaving the door open for her and greeting her with some idea of complicity whenever he met her on the staircase. By contrast, he was horrified by the attitude of the patient who preceded him on another day of the week; that person would bang the door shut as he left and pretend not to see him sitting by the window ledge waiting his turn. A memory that did not seem to know how to be a memory was clearly present in such trivial and apparently insignificant details.

The idea that he had to write down what had taken place occurred to him several times. He thought about it that evening, the next morning, the following evening ... but in fact did nothing about it. It was as though, in this early and fragile phase of the lifting of repression, forgetting was the natural course that his mind would take, whereas remembering would be the expression of strategy and hard work. The patient talked again about his relationship with his grandparents. What struck him at that point was the attitude of his grandfather. How could that man, the patient suddenly wondered, who weighed almost sixteen stone and was nearly six feet tall, have slept in the small bed, giving up his rightful place beside the patient's grandmother? That memory was filled out by something else that he recalled: since the bedroom was unheated, his grandfather used to place a hot brick between the sheets of the double bed a few hours before they went to bed. So, not only did his grandfather give up his own rightful place, he even warmed it up for his grandson! Once the patient got over the stupefaction that that discovery brought in its wake, he felt overwhelmingly sad. Had he charmed his grandmother to such a degree that, in unison, they had evicted the grandfather? I pointed out that shutting the door on the following patient would seem to have something to do with an eviction; I suggested that, at the end of the previous session, when he thought of her he was also perhaps thinking about his grandfather.

At that point the patient fell silent for a considerable length of time, something quite unusual for him. What I had just said made him think of M, his sister. She had been physically abused by her partners, and he himself had been violent towards all the women with whom he had been in a relationship. He had never before seen any connection between all that. His sister was very close to their mother, while he was very distant. Above all, he found himself thinking about something and wondering why it had never come into his mind before: when he was born, because he suffered from a particular somatic disorder, his mother had asked her own mother, the patient's grandmother, to look after him, but when his sister M was born, just eleven months later, mother kept her and looked after her. Pain and the feeling of having been treated unfairly swept over him. I commented that perhaps he had closed the door after that other session so that I would not be able to keep the next patient, a woman, and look after her. Thinking about that other patient probably had something to do with his sister, and thinking about me had to do with his mother. That fragment of recollection was a great relief to him.

These apparently insignificant details of an ordinary analysis are, none the less, valuable in that they illustrate the way in which the work of remembering operates. It cathects the slightest incident that is inherent in the situation, conferring on it a certain memory value by turning it into a representative of unconscious material wiped out by infantile amnesia. To my intervention signifying the end of the session—completely neutral and forced on me by my very role as psychoanalyst—the patient ascribed the meaning of an undesirable and disputed paternal authority. To the presence of the woman patient about whom, objectively speaking, he knew nothing, he ascribed an Oedipal value, that of representing his sister, the hated rival in the vital attachment linking him to his mother. Unconscious fragments from his childhood past were thus projected on to the minutiae of his present life. The mechanism of projection is an important ingredient of remembering: indeed, it is (perhaps frequently) its initiatory phase; in the face that a new object presents to me, what I see first of all is an old object, one that I have in my memory.

None the less, this first level of analysis in terms of fantasy content does not do justice fully to the work of remembering, which, at the same time, reconstructs the real historical context thanks to which those fantasies are built. Significant people from one's childhood, their emotional importance preserved in unconscious memory like photographic negatives, are gradually revealed as they become personified in the analyst or in some other person having a present-day connection with him or her.

Think of the way in which we decipher a photograph. After identifying the people represented in it, easily and quickly done, we begin, more slowly and with more difficulty, to study the place where it was taken, observing the clothes that the people are wearing and the surrounding furniture, until gradually our memory of the whole circumstances (a first communion, a family reunion . . .) becomes clearer and clearer. Similarly, the reappearance in that patient's transference of the image of his grandmother, then of that of his grandfather, recalled to his mind the painful memory of the passionate relationship that he knew—and did not know—that he had formed with them.

A detailed reconstruction of the history of events that lie at the core of traumatic memories—Freud gave several examples of this both in the case of *Dora* and in that of the *Wolf-man*—corresponds to what, in psychoanalytic theory, is called the *lifting of infantile amnesia*. The recognition of the Oedipal objects that go to make up unconscious fantasy corresponds to what is called the *lifting of repression*. It is worthwhile keeping these two processes separate, because the successful accomplishment of the one depends on that of the other. That is the conclusion that Freud came to in the paper he devoted to a thorough study of remembering, "A child is being beaten". The reconstruction (by the analyst) of the second phase of that fantasy ("I am being beaten by my father, therefore he loves me"), which reveals the incestuous drive impulses at stake underlying the historical event, becomes possible only through the active work of remembering, which is different from the work of interpretation revealing the objects involved. This is how Freud puts it:

> Strictly considered . . . analytic work deserves to be recognized as genuine psycho-analysis only when it has succeeded in removing the amnesia which conceals from the adult his knowledge of his childhood from its beginning (that is, from about the second to the fifth year). This cannot be said among analysts too emphatically or repeated too often. . . . The emphasis which is laid here upon the importance of the earliest experiences does not imply any underestimation of the influence of later ones. But the later impressions of life speak loudly enough through the mouth of the patient, while it

> is the physician who has to raise his voice on behalf of the claims of childhood. [1919e, pp. 183–184]

What is striking about the clinical illustration I have just given is the fact that the initial recollection involving a fairly late stage in childhood—indeed, almost pre-adolescence—was immediately followed by one which had to do with a much earlier recollection that went back to the patient's first year of life. Can the latter justifiably be called "remembering"?

"Being abandoned" by his mother—which is what the patient's narrative was all about—did not belong to his own memory; he learnt about it from the family saga, as it were. That event took place at a time when, as a young infant, the patient had no words at his disposal, so that, no matter how brutal the experience might have been, it could only have given rise to sensory traces that in themselves were meaningless; they had no mental substance other than through their connection with the verbal memory involving the second, later, event. Therefore, it would be more exact to say that the first instance of remembering, which emerged dressed in words, both carried within itself and was brought to life by a recollection that was not a memory *stricto sensu* and had no linguistic content, something like a flame that points to the existence of the glowing embers which give it energy without actually revealing them. This, then, was not a case of *remembering*, but of a *reconstructed memory*. I would like to emphasize the fact that this "reconstruction" took place through the mediation of language and the linguistic processes of free association and interpretation. These operations enable us to read the foundations that are invisible to the naked eye and inaudible to the immediate ear of unconscious memory; those foundations are what carry the tragic potential of that recollection. To the sensory traces constituting the biological memory that archives the traumatic experience which the outside world has inflicted on the individual, there is thereafter added the memory function of language on which is based the kind of memory that is specific to human beings. As Pierre Fédida (1995) put it: "I cannot say what happened, but I can create a *locus* for what happened. That *locus* is psychotherapy". He went on to say,

> Did the event itself actually occur? What is important is the narrative in which it will exist. What is important is not the traumatic

> event that took place when the person was two years old, but the setting up of a mental *locus* in which that traumatic event can occur in the course of the analysis. [*ibid.*, pp. 81–92]

In psychotherapy, the *locus* is, of course, that of language, discourse, narrative, and dream reporting.

A detour via a particular aspect of pictorial expression will further our understanding of the strange way in which the human mind works in this respect. The X-ray examination of Velasquez's well-known painting, *Las Meninas*, shows that, initially, it represented a formal ceremony: the French ambassador, on the right-hand side of the painting, was presenting to the *infanta* Marie-Thérèse her marriage contract with the heir to the French throne. After a younger brother was born, the *infanta* lost her rank as the future queen, and the marriage contract was annulled. Velasquez thereafter changed his painting, putting a sketch of himself where the ambassador had been. Knowledge of this reworking, via this highly indirect detour that technique offers, is crucial for our understanding of the painting: it explains its tragic dimension, summed up in the way in which the *infanta* screws up her face, contrasting with the harmonious mildness of the rest of the painting. The loneliness of that child, condemned to be queen although she would never actually become one, the distress of the *infanta* faced with a destiny that is both frustrating and overwhelming, that is what the painting enables us to understand indirectly, that is why we find it so moving. Even though the retouches are invisible, they give substance to the painting.

In the work of remembering that takes place in the course of an analysis, language is the *technè* that indirectly captures and reveals the negative dimension of memory, or, more precisely, the negative and infinite memory that continually nourishes recollection.

We must, therefore, draw the conclusion that the way in which experiences become part of individuals and turn them into subjects of their own history follows a strict course that the work of analysis must discover and to which it must submit in order to be able to go backwards in time. Memory is not passive, it is not a mere storehouse for the relationship between the individual and the real world, it is selective. It can identify in events that temporarily destabilize psychic equilibrium—and thus may be experienced as traumatic—the underlying drive-related features common to all. On that basis, it condenses them, thereby allowing them to represent one another, so that, in the memory chain thus set up, the most recent (which, thanks to its link with language, can be recalled) is given the task of slowly but surely evoking the oldest, which is in itself barred from memory.

In the example I reported, two aspects were common to the patient's later recollection concerning his grandparents and the constructed memory about being "abandoned" by his mother: first, the idea of separation that could only at a later date be expressed through a *verbal signifier* (before, it was expressed as a *sensory signifier*); and, second, the regressive hyper-sexualization that we find both in object deprivation and when relationships become too close and incestuous.

Although interpretations are, of course, something that the analyst suggests, they cannot be dissociated from what the analysand says: that is, after all, their very foundation. The effect is not simply to reveal the disguised memory in the perception currently being put forward, but also to compel that memory to give up the Oedipal satisfaction that is secretly part of it. "Rediscovering" his grandmaternal and maternal objects in the perception that he had constructed of the analyst as a person inevitably leads the patient to relinquish them, to "lose" them. The *work of the analysis*—and André Green has put forward some very convincing arguments in this respect—takes *dream-work* as its first model, and, as its second, the *work of mourning*. In analysis, we subject ourselves to the ordeal of giving up the dearest objects of our childhood that we had transgressively preserved.

On the other hand, the device, the art, that controls infantile amnesia seeks to preserve our precious Oedipal objects. It spares us from having to undertake the work of mourning that leaves its painful mark on the love-life of childhood. The violence of interpretation, to use Piera Aulagnier's powerful, but none the less exact, term, has to do with the instruction it carries within itself like a sharp point: to forget the object which repression mischievously keeps alive, on the pretence of abolishing it. Remembering works at forgetting, not in the usual sense of the word, that of a lazy and negligent act that avoids anything unpleasant, but in terms of taking a decision consisting in acknowledging the fact that what exists no longer does, and of what did exist all that remains is the memory of, and nostalgia for, it.

Only works of literature can adequately depict that quality of the human mind which consists in making use of memory in a supremely skilful way to combat forgetting. I mention this with the precise aim of drawing a parallel between the economics of memory in the course of an analysis and that found in novels. The *lost time*, in search of which Marcel Proust set out, amounts to a time that may perhaps be found again, one that the magic of the story saves from ever becoming a bygone past. The extracts that follow are taken from Proust's "The captive".

On that particular day, Albertine had refused Proust's offer to accompany her to the Verdurins'. Suspecting that she is being unfaithful to him, he advises her to go to "the splendid charity show at the Trocadero". Albertine listens to that advice "with a sorrowful air". The words "charity show" sound somewhat strange to modern ears and—like the signifier "murder" in the dream of the analysand that I mentioned earlier—is unnecessary for the development of the narrative: it is superfluous. That incongruity bears witness to its function as an anacoluthon, a figure of speech that, according to Roland Barthes, mediates between the manifest current discourse (i.e., the narrative) and an inner discourse, underlying the former and devoted to the expression of the internal world, its fantasies and its memory. It indicates that the author, although apparently describing what the character is saying or doing, is in fact talking about and to himself. Proust then goes on to say:

> I began to be harsh with her as at Balbec, at the time of my first jealousy. Her face reflected a disappointment, and I employed, to reproach my mistress, the same arguments that had been so often advanced against myself by my parents when I was little, and had appeared unintelligent and cruel to my misunderstood childhood. [pp. 107–108]

The description of that passionate scene fills the following few pages, which are deeply moving. Proust both narrates the scene as it actually took place, with his scolding and Albertine's despondency, and includes his own thoughts in which he activates, for his own private use, the memory of those near and dear to him who have departed. The first of these is his mother. This is what Proust says of her: "These words—a great part of what we say being no more than a recitation from memory—I had heard spoken, all of them, by my mother . . .". He goes on to speak of his grandmother: "Her severity towards myself was deliberate on her part and indeed cost her a serious effort", and then of his father: "Perhaps in my father himself his coldness was but an external aspect of his sensibility".

At no point, however, even although these evocations are expressed clearly and precisely, does Proust realize that Albertine is not his mother, or his grandmother, or his father. That indeed is the advantage of romantic fiction: to call upon memory and take pleasure in its capacity to comfort and in its charms, without touching its integrity or virginity, without compelling it to remember. In this dialogue between lovers which mingles the relentless moments of past and present, of childhood and adulthood, and in which the participants are completely open about their feelings, Albertine is both the child that Marcel once was and the parents to whom he can at long last throw back the hurtful words that they had said to him in earlier times. Albertine is no more than the excuse that enables Proust to restore those bygone days; they are reincarnated, almost in a hallucinatory manner, through her.

Memory is a delicate thing. Preserving the past helps us to live, but it can also make us ill. Between these two alternatives, only the individual can decide. I shall let Proust sum up what I have been trying to say here. In order to reconcile himself with (and resign himself to) his "repetition compulsion", this is what he writes; this passage, to my mind, is of great value both because of its beauty as literature and because of its elective affinity with psychoanalytic thinking:

> No doubt, as each of us is obliged to continue in himself the life of his forebears, the balanced, cynical man who did not exist in me at the start [i.e., of this scene] had joined forces with the sensitive one, and it was natural that I should become in my turn what my parents had been to me. What is more, at the moment when this new personality took shape in me, he found his language ready made in the memory of the speeches, ironical and scolding, that had been addressed to me, that I must now address to other people,

> and which came so naturally to my lips, whether I evoked them by mimicry and association of memories, or because the delicate and mysterious enchantments of the reproductive power had traced in me unawares, as upon the leaf of a plant, the same intonations, the same gestures, the same attitudes as had been adopted by the people from whom I sprang. [pp. 107–108]

Who can blame Proust for being so fond of his melancholy when, through the magic spell of his writing, he brings his readers into such absolute complicity with him?

References

Benveniste, E. (1966). Le langage et l'expérience humaine (Language and human experience). In: E. Benveniste et al. (Eds.), *Problèmes du langage* (Problems of language). Paris: Gallimard.

Fédida, P. (1995). L'epos—le site (Epos—the site). In: *Le site de l'étranger: la situation psychanalytique* (pp. 81–92) (The site of the stranger: the psychoanalytic situation). Paris: Presses Universitaires de France.

Freud, S. (1900a). *The Interpretation of Dreams. S.E.*, 4–5. London: Hogarth.

Freud, S. (1915). Letter from Sigmund Freud to Lou Andreas-Salomé, 30 July, 1915. In: E. L. Freud (Ed.), T. Stern & J. Stern (Trans.). *Letters of Sigmund Freud 1873–1939*. London: Hogarth, 1961.

Freud, S. (1915e). The unconscious. *S.E., 14*: 159–216. London: Hogarth.

Freud, S. (1919e). A child is being beaten. *S.E., 17*: 175–204. London: Hogarth.

Freud, S. (1920g). *Beyond the Pleasure Principle. S.E., 18*: 7–64. London: Hogarth.

Freud, S. (1922b). Some neurotic mechanisms in jealousy, paranoia and homosexuality. *S.E., 18*: 221–232. London: Hogarth.

Freud, S. (1937c). Analysis terminable and interminable. *S.E., 23*: 209–254. London: Hogarth.

Green, A. (1993). *Le travail du négatif*. Paris: Éditions de Minuit (English edition: *The Work of the Negative*. A. Weller [Trans.]. London: Free Association, 1999).

Green, A. (2000). *Le temps éclaté* (Fragmented time). Paris: Éditions de Minuit.

Picoche, J. (1971). *Nouveau dictionnaire étymologique du français* (New Etymological Dictionary of the French Language). Paris: Hachette-Tchou.

Pontalis, J. B. (1997). *Ce temps qui ne passe pas*. Paris: Gallimard.

Proust, M. (1923). *La prisonnière* (The captive). In: *À la recherche du temps perdu*. Paris: Gallimard. (English editions: *In Search of Lost Time*, C. K. Scott-Moncrieff [Trans.]/*Remembrance of Things Past*, D. J. Enright [Trans.].

CHAPTER FIVE

The time of the past, the time of the right moment

Janine Puget

Stagnant time

Some comments we hear in everyday conversations reflect a belief/conviction that the past (whether one's own or historical past) was better or easier than the present. "Everything was easier," said a patient. "There were no computers, traffic wasn't so bad . . ." "It's true," I answered, "but there was no penicillin, and people died younger." A remark like the patient's contains the idea that the present should be like the past, and is a sign that the present poses an obstacle. Regardless of its specific, contextual meaning, it betrays a difficulty and a dissonance: a difficulty in thinking of the present and in the present and in doing something with it, and a lack of harmony between present and past. Present and past might even be separated by an interface, and progress by leaps and bounds.

An Argentine bolero (singer), a source of popular wisdom, speaks about this type of nostalgia: "There is no worse yearning than longing for what has never, ever happened" (Sabina, 1990). Another songwriter mentions being "trapped in my yearning for what couldn't be" (Gonzalez, 1996). Such longing speaks of an imaginary past, a bygone past, or one that never happened. Perhaps this past provides comfort in that it is deprived of the uncertainty of present-day vicissitudes, and does not demand great effort. It is recalled, but not experienced, in the here-and-now, and it removes us from the complexities of the present.

Is there such a thing as a pure present? Or does the present always bear qualities of the past? While the present invested with the past plays a role in most psychoanalytic theoretical presuppositions, the pure present puts these presuppositions to the test. It is valid to conceive of a present that comprises an interweaving of models and values whereby past and present influence each other. Yet, we must also take into account (and this is our greatest challenge) both an evental[1] present that disrupts linear temporality and the associative chain, and a new, unpredictable present, the present of the encounters between two or more subjects. These presents open a gap in the order of things, and surprise us.

Since they do not provide the distance required for reflection and verbalization, both the evental present and the pure relational present render speech, thought, and narrative difficult. The present-past, by contrast, may be discussed by means of memories. Both the evental present and the relational present instil a lived experience that is grasped through events themselves, but cannot be narrated; it is experiential. These presents shake illusorily solid and certain subjective positions and, being always new, pose their own obstacles. They disrupt essential subjective positions and dislodge subjects from solidly constituted places that fit into unconscious family structures and already built social subjective structures. (By solid structures, I mean those that assign traditional places within the family structure, that is, father–mother–child, or simply those that correspond to the social structure that harbours us as social subjects inscribed in a group–country.) They upset charted courses, give way to the uncertainty inherent to life, inaugurate an unpredictable future, and sometimes highlight contradictions, or even enigmas. It is likely that when subjects face this type of situation, some form of discomfort or anxiety will be activated. In other circumstances, creativity, the generation of new ideas in the here-and-now, is either blocked or fully developed.

Defences against the relational present emerge, for instance, in remarks that disavow the new and surprising aspects of any encounter: "I know you, you're always the same . . ." This kind of reaction halts the subjects' becoming, or the unfolding of their subjectivity in the present. Other comments, such as, "the same thing happens to us every time", or "this relationship is going nowhere", are also signs of the subjects' difficulty in experiencing the present, which implies an inability to draw on the creative power of a relationship. Such inability might stem, among other things, from a mistaken assessment of the situation due to the subjects lacking categories to approach the pure present and the evental past. As a consequence, they search for similarities between two contextually and epochally different moments. In politically complex times, people tend to see the present as a premonition of the return of the past. The same thing happens when a couple or a family experiences a crisis and sees it as the repetition of a past situation. The denouement is thus deprived of its novelty. The narrative bears expressions such as "always", "never", "I already knew that", and so on. Yet, evental present resists being thought of in terms of binary categories, and inaugurates a sometimes enigmatic subjective space.

Both in everyday exchanges and in psychoanalytic practice, we hear remarks that evince a longing for that past that was better and easier to understand. "I came because I feel bad, but I was fine before [an ungraspable before]. These things used not to happen to me. I'd like to *go back* to feeling that way." The verb "go back" is a sign of a nostalgia for the past, and it rejects the search for new paths. Couples complain of having lost a "before", perhaps the time of infatuation, or another time when the encounter between them did not create interference. Families long for those moments when "the family was together", whereas now "each goes his own way . . ."

Even when it provides new elements that could make our lives easier, the here-and-now appears as an obstacle. Moreover, if we take into account that subjectivity is also defined by the ways in which we occupy positions in the broader socio-economic context, events that lead to the questioning or invalidating of positions of certainty doubtlessly confront us with a frightening future. Psychoanalysts are not spared such difficulties. Sometimes, we expect yesterday's analysand, and miss the chance to listen to and create alongside today's. Thus, we preclude coming into contact with the discontinuity between past and present, and avoid the surprise of the encounter.

Personal relationships: complementarity and supplementarity

When we analyse personal relationships as transformations of the first, parent–child relationships, it is likely that complementarity will play a key role. If, by contrast, we attribute a specificity to the encounter between two that hinges on the role played by each participant's otherness, the identity-based notion loses prominence, and a greater challenge is posed. It is a matter of subjects' welcoming into their minds and lives what exceeds them; that is, what is foreign to them. Exchanges might not only fill a lack; they might also provide a supplement by generating something that had not been thought or experienced before.

The encounter is endowed with the potential to create something common to all participants, and to produce something new while respecting the space in between. This is an empty space that will remain so, but that is, none the less, the necessary condition for the constitution of the subjects who forge that particular link. The common or shared aspects of relationships are not construed on the basis of similarities, but of differences. The space in between demands from the participants their tolerance of emptiness. The relationship acquires meaning to the extent that a work is carried out by the subjects involved. The empty space must not disappear.

We might think, then, that even though the work the subjects carry out together is triggered by difference, it is this emptiness that is common to them, as Lewkowicz (2004) suggests in his modification of the schema set forth by Badiou (1999). Laclau (2008) points out that for Badiou, the empty space in between is inherent to the situation and traverses it. It is indiscernible and bears nothing that the situation can recognize. In a love relationship, love deprives the emptiness of its worrisome quality. Laclau goes even further when he discusses Badiou's analysis of St Paul's discourse (Badiou, 2003). He suggests that when something is named that "does not correspond with anything that can be represented within the appraisable dimension of the situation, we might think of emptiness as a signifier without a signified" (p. 76).

Each of these theoretical presuppositions refers to the idea that the space between two subjects bears a relational potential, and that it is what enables the construction of something common to both participants. In many theoretical formulations, by contrast, common aspects are conceived of as similarities rather than as those elements that, being different, demand a specific work.

Other authors ponder the constitution of community, what is done together. Doing entails an act; it means working in the here-and-now. In Arendt's (1958) terms, doing renders subjects human, and according to Agamben (2000), it depends on "the experience of thought" as an "experience of a common power" (p. 20). This way of doing might be homologized to the production of meaning. One of its outcomes is the construction of communities that place the emphasis on what they mean by "common". I base this analysis on the ideas of authors such as Lévi-Strauss (1966), Agamben (2000), and Espósito (1998), who study the complexity of exchanges and their various modes. I will only linger here on Espósito's analysis of the word "community". Starting from its etymology, this author maintains that the word "common" is semantically complex. The particle *munus* refers to meanings that include the ideas of duty, gift, and obligation. Espósito's analysis stresses the difference between *munus* and *donum*, which correspond to different modes of exchange. None can be rejected. Both start a process of giving and receiving whereby we give because not giving is impossible. We give expecting to receive; we receive without having to give back. Giving deprives the subject of something, and does not imply reciprocity. Moreover, the other *must* accept the gift (*ibid.*).

There are many ways to create common spaces, and all of them force participants to move away from solid identity positions. This movement, based on a work that takes difference into account, poses challenges that become apparent, for instance, in the subjects' search for factual agreements that curtail the relational potential of radical difference. Building commonality only depends, as Benjamin also noted, on the setting in motion of a quality that he called "communicability" (Agamben, 1993, p. 21), and the exchanges inherent to relational life (I refer to exchanges among subjects with this term in order to go beyond the various vicissitudes of projective and introjective identifications) give rise to new productions. The proposals advanced by Lévi-Strauss, Agamben, and Espósito share the idea that exchange demands both tolerance toward difference and the performance of an act that bears its own rules.

Becoming

My research has gradually turned to the study of the subjects experiencing and forging relationships within a temporality that comprises both their gradually becoming with others throughout the course of their day-to-day experiences, and evental present. Remarks such as, "But if I get along with my friends, I don't know why X doesn't realize who I am", denote a way of thinking of oneself based on an essential, and, hence, identity-orientated, perspective. Consequently, they do not give rise to a process of becoming, or to subjectivizing productions that change with each relationship. For a long time, I thought that such variations could be read in terms of dissociations, or of instances of splitting of the ego. Even though, at times, this approach might still be valid, it is worth considering that encounters between two or more othernesses open gaps. They interfere with the identity-based conception of human life and place the subject/s in a pure present.

A couple came to see me and said to me, "We'll tell you the history of our relationship so that you can understand what's happening to us now." In this case, the causal hypothesis invests the present with a repetitive feature that avoids every aspect of this present that cannot be properly pondered or experienced. The patients believe illusorily that if the psychoanalyst understands the past, she will understand what is going on today.

Present and future

Taking into account the gaps between past and present guides the work towards the definition of temporalities where present and future are no longer mere repetition and transformation of something that has already been experienced, but pure novelty that is renewed daily. A patient told me that she could not listen to her partner's praises because she kept thinking of his previous unfaithfulness. Every present situation would remove her from that painful moment to which she remained adhered in a sort of narcissistic delight. The new "present" introduced a novelty that did not articulate with her memories.

Considering present temporality from the perspective I am suggesting here takes us away from the exclusive representational world, and brings us closer to the presentational world (Puget, 2003). I will not dwell on this topic, but I would like to point out that I base my analysis on the presupposition (suggested by Nancy [1986], among others) that the need to differentiate between an order of representation and an order of presentation stems from the confirmation that a crisis of representation has occurred. The term *presentation* refers to an encounter between two or more subjects that sets in motion the *judgement of presence* (Berenstein, 2001; Puget, 2002). Through this type of judgement, subjects become aware of the presence of the other. The judgement of presence joins the judgements of existence and attribution, whereby subjects confirm their unique existence and their own qualities.

The experience of a present without traces inaugurates a new future. Empty spaces, breaks, and ruptures must be tolerated. As events take place, empty spaces continue to be created in the manner of mirages, which take different shapes depending on who occupies the edges. Such shapes stem from doing with an/other and among others, and are fleeting and highly uncertain.

Biological conception vs. relational subjective conception

I wonder to what extent certain working hypotheses, set forth in a great number of psychoanalytic theoretical corpora, are not fostering an idealization of the past. Such idealization certainly contains and promotes a melancholy component—what I have called the nostalgia for a past that could not be—and is also an inexhaustible source of explanatory models. Undoubtedly, in its inceptions, psychoanalysis had to focus its attention on identity and on the constitution of a singular psychic apparatus. Yet, this move set aside the analysis of the complexity of relationships between two othernesses and with the social world, relationships that hinge on each subject's irreducible difference.

Most theoretical models that aim to explain psychic functioning are based on a biological conception whereby the newborn is helpless and defenceless. From this perspective, parental figures become complementary objects that satisfy the infant's needs and help it find its bearings in the world, thus instituting it as a human subject. Infants are, in a way, formed–deformed to become subjects, as Aulagnier and others have suggested. These authors see the relationship between the baby and its parental figures as instituting, and being the bearers of, past values that are gradually assimilated and transformed by the child. Notwithstanding the validity of this description, it should not prevent us from considering that relationships produce subjectivities and new vicissitudes through the work demanded by radical difference, the unique components of each relationship.

Which working hypotheses anchor the ego, the subject, families, couples, and generations in the past?

The representational world facilitates our understanding of the psychic functioning of single subjects in their relationship with the external world, in so far as the latter enables them to constitute themselves as subjects of their internal world. To do so, objects or people in the external world with which subjects come into contact—the parents being paradigmatic—become objects of the drive, objects of desire, with the consequent denouements. These objects gradually become subjects/objects that are incorporated into the subjects' world to populate it and to promote dynamic exchanges among the various agencies that comprise what we call the psychic world. Theories that focus on the representational world—from Freud, to Klein, Laplanche, Aulganier, and so on—usually include a history based both on genetic continuity and on the reversibility of the arrow of time. I will review here some presuppositions that underlie the various theoretical models that, despite their differences, endorse this perspective.

The theoretical presupposition of the *trauma* (of birth) as the *origin* of the psyche—that is, of individuals' life as human beings—suggests that ideas give shape to sensations, and vice versa. The notion of origin bears great explanatory power, whence its appeal. Origin is associated both with the original helplessness and the trauma it produces, and with the first object relations, the first contact with a specific other who provides care and complementarity to the handicapped infant. It guides the analysis towards the search for historical data that facilitate the reconstruction of the past in order to dig it up and turn it into memory and also, on occasion, modify its negative qualities. At the same time, origin also becomes a force of attraction that checks the process of becoming.

The *question of origin*: from a methodological point of view, it might be useful to create or establish an origin or point of departure based on which we may formulate explanatory models of some sort. Yet, being methodologically necessary does not mean bearing an empirical or epistemological basis. This origin has no ontological value; it simply prevents infinite regression. The origin of a psyche, and hence its constituting mark, has, therefore, a mythical value. The myth suggests a scene with its own dynamic, and provides an explanatory tool. From the myth stem codes and signs that organize a new relationship between two or more subjects within a cyclic temporality.

The *drive*, privileged origin of the subject's becoming, is also viewed as the basis of psychic activity. The drive-desire is an inexhaustible force that promotes the search for an impossible satisfaction. And, given that desire remains unsatisfied by definition, it stimulates, through successive transformations, the generation of the idea of the future, of a moving forward that takes subjects away from their dissatisfaction. Yet, this process is twofold. On the one hand, it drives subjects forward; on the other, it leads them illusorily to seek to recover a past satisfaction. I already gave the example of a couple that, in view of current difficulties, longs for a time of infatuation that they remember as conflictless. And they are right in a sense, because it was the time when each subject's otherness was concealed by the mutual strengthening of narcissistic cross-identifications.

Regression necessarily accompanies this way of thinking. Subjects are attributed a particular ability, that of reverting the arrow of time. In this way, they can defensively regress, albeit imperfectly, to previous states. This process is ruled by a circular time that gives rise to myths, fantasies, and illusions. None the less, as many philosophers have pointed out (Nancy, among others), if "nothing is reencountered in history, neither do we return to anything, either God or values" and subjectivity entails a quality of encounter, we must differentiate between reencounter and encounter, both theoretically and clinically. If an inscription (mark or trace), an invariant, an identity trait, or an illusion is recovered,[2] it means that it was but a momentary loss sustained by various psychic occurrences. Ruptures are interruptions, for meaning can be recovered. Granting them a sense of loss or lack creates a type of discontinuity whereby experiences and events are articulated and inscribed on a continuous background. It is, therefore, possible to think that what was lost can be recovered, for it is inscribed in a successive history of marks and events.

Fixation points suggest that, during the evolution of human subjects, knots are produced that represent an arrest of development and act as an attracting pole. A difficulty or a traumatic situation checks the progress that subjects or relationships need in order to acquire new qualities. Thus, subjects remain in a different temporality, that of a past that permeates the present where they are anchored, trapped.

We also need to review the role of *absence* and *lack* as the only forces capable of generating thoughts.

Evolutionary theories contain the idea of the reversibility of time. In this context, I attribute particular significance to the principle of genetic continuity. Some authors (Isaacs, Rivière, Heimann) see it as a "concrete instrument of knowledge" and one of the "essential aspects of the work of psycho-analysis" (Isaacs, 1948, p. 78). This principle establishes that "any given phase develops by degrees out of preceding phases in a way which can be ascertained both in general outline and in specific detail" (*ibid.*, p. 77). Furthermore, every fact is part of "a developing series" and a manifestation "of a process of growth" (*ibid.*, p. 78). Isaacs claims that

> in his phantasies towards the analyst, the patient is back in his earliest days, and to follow these phantasies in their context and understand them in detail is to gain solid knowledge of what actually went on in his mind as an infant. [*ibid.*, p. 80]

These statements, which partly coincide with Freud's criteria of genetic continuity, imply that we can extrapolate a hypothesis produced in the here-and-now to the very first mental processes. From this perspective, evolution would not happen in leaps, but would be a continuous process (Lores Arnaiz & Puget, 1981, Puget, 1994).

In this same scientific context, *transference*, in as much as it enables us to consider the here-and-now as part of the there-and-then, would provide significant knowledge while reinforcing the idea of time's reversibility, or at least of the prevalence of circular time. Moreover, in another of these authors' postulations, transference also eliminates the present analyst's novelty, alterity, and exteriority. Interfering presence and otherness are cancelled by means of the mechanisms that set in motion introjective and projective identifications of the analysand with the analyst's figure.

These hypotheses are not the only ones that psychoanalysis formulated to support the idea of circular time and the reversal of the arrow of time, but they are probably the most important ones.

Is there room for current events and new experiences?

The above-mentioned hypotheses, which carry significant weight due to their explanatory power, may sometimes operate defensively. When, for some reason, patients attempt to avoid the suffering caused by taking into account the unpredictable and new aspects of the present, approaching the clinical material from such theoretical framework refers them back to the past instead of confronting them with these elements.

Since I am suggesting that every encounter necessarily surprises subjects and dislodges them from solid identificatory positions, we also need to rethink how we address our analysands' history and historicization processes. Through the history narrated by the analysand, we may sometimes give meaning to fragments of the manifest content, or make predictions. At the same time, we may also think that what becomes present in the here-and-now of the analysand's history contributes only one explanatory element. In my view, a different concern leads analyst and analysand to build both the history of the present and a new history based on that present (Puget, 2006a,b).

I have often found myself resorting to the figures of my patients' infantile past, a past we had already visited in different ways, to have the patients say to me, "But that can't still be valid . . ." And it turns out they are right. Something must have been happening with my understanding of the material that emerged in the sessions for me to continue to seek support on a past that was losing its explanatory power. At other times, I have been led to question my own thinking by patients who, having had several analytic experiences, came to me saying that they did not want to tell their story again, and so they thought they did not want to be in analysis, but to do something they called "therapy". Even though we might see this attitude as a form of resistance, I also think of it as a search for meaning that might lead to creative encounters. This approach has opened the way for working on the pure present. Focusing on the past might then be a sign of a theoretical obstacle.

Does psychoanalysis mean dealing with the present?

Tackling the present's history entails a particular way of looking at what takes place in the session. Analysts must recognize how an experience is lived, as well as the specific elements of a relationship between two subjects (patient and analyst). They must also understand what is involved in talking about this experience as participants in the dialogue instead of as observers (as witnesses capable of communicating their view of what is taking place). Sometimes, such an attitude could be confused with a countertransference confession. None the less, it is but an intervention that places analysts as subjects of the relationship rather than simply as subjects–objects of the transference.

Analysts' interventions might be somewhat descriptive, providing a view from a vertex different from the analysand's. Other interventions might open a gap in the analysand's approach. When seeing the analyst as an object of the transference, analysands could experience this incongruity between the two versions as the analyst's lack of understanding. Comments such as, "You don't understand me . . . I meant this and that", or "I don't agree with you", will follow. Analysts, in turn, might convey the idea that they are seeing this situation from an "other's" place, which might not coincide with the patient's experience of it. This lack of agreement should lead to a new idea, a new way of thinking. Analysands strive to impose already known meanings; analysts, to open a gap in the identificatory excesses favoured by transference theory.

The complexity of this interaction is due to the fact that analysts undergo two different experiences with their analysands: that of transference–countertransference, and that of intersubjectivity. Experiencing intersubjectivity entails letting oneself be carried away by the becoming of the relationship and establishing a dialogue that takes place in the fluidity of the encounter. Perhaps this attitude entails what has been described as suspended attention, or, in Bion's terms, being without memory or desire. In this analytic present we listen, answer, and speak according to what emerges in the moment of the experience. These aspects of the analyst–patient relationship that transcend the transference–countertransference dimension led Berenstein and me to suggest a term, interference, to describe the effects of presence on the analytic process. Interference names that which exceeds subjects and their internal–external objects (Berenstein, 2004; Puget, 2001).

There is another time available to analysts, that of reflection: a momentary interruption of dialogue so as to step aside and talk about the experience, about what is taking place. This is a different time from the pure present. It is a time when we consider a recent past, something that has already happened. We are dealing here with two presents, perhaps with two pasts. One of the pasts is that of the analysand's history and of the history of the analytic relationship. The other is the recent past, which will become present if we are able to instate a time of reflection.

The dialogue that unfolds in the analytic space has unique features. It is different from those taking place in other spaces. It stems from a specific device: analytic time. None the less, it can be transferred to other spaces. Some remarks illustrate such a transfer, for instance, when analysands say about their conversations with their partner, "We went to the café because at home we can't talk." Of course they talk at home. Yet, they express ideas that sometimes follow the rules of administrative or financial exchanges, where the time of the present and the time of reflection about that present (the time when partners meet to listen and to be listened to, to talk to each other as othernesses, and to create what is common to both) are absent. This talk they try to produce or create outside the session might be one of the effects of shifting the products of the analytic session to other relationships. We should not consider this behaviour as acting out, but as the ability to make use of a new tool, that of being with an/other while incorporating his or her otherness and exteriority.

Experience and the right moment: kairos

Then, how do we define this space, the space of analysis, which produces a unique experience? And how do we make room both for the past of the present, which permeates experiences, and for the present that creates a new past?

For now, I would like to focus on the "right time" and define the experience of the moment as typical of the present situation. Each of us has had the experience of listening–receiving or uttering the right word at the right time, and along with it, the experience of an encounter with an/other that defined the course of a relationship. Sometimes, couples try to capture the "right time" they value as the origin of their connection. That is when this moment, the moment of the encounter that generates decisions, becomes anniversary. Sometimes they say, "We celebrate the day we met." That was the right moment. This moment happens within a favourable situation. It is the right moment because it can bring about change, discontinuity, and because it combines various conditions (affective, intellectual, emotional, and so on).

In other cases, the right time is not when we listen to a certain word or respond to a gaze, but when we connect with a book, a work of art, or a landscape that produces an epiphany and a change of perspective. We may define the right moment and the consequent epiphany by means of Benjamin's concept of "aura", that is, "the unique phenomenon of a distance, however close it may be" (Benjamin, 1969, pp. 222–223). Benjamin frames this definition in the context of

> the formulation of the cult value of the work of art in categories of space and time perception. Distance is the opposite of closeness. The essentially distant object is the unapproachable one. Unapproachability is indeed a major quality of the cult image. True to its nature, it remains "distant, however close it may be". [*ibid.*, p. 245]

In the same way, the right moment vanishes when we try to grasp it, thus retaining its mystery. Benjamin assimilates the notion of aura to the authenticity of a work of art, for reproductions are not endowed with this quality. The aura is also something indefinable that anticipates an event and might even facilitate its predictability.

Yet, it seems that the mind strives to grant materiality to this evanescent state, and somehow situates it in time and space. Something of this materiality is present in dates that become significant, such as birthdays or anniversaries of a happy or sad moment, dates that may be personal, relational, or social. It would be impossible to translate into words the myriad feelings and emotions contained in these dates. They certainly bear their own authenticity. Then, why not think that anniversaries unite both presents? Anniversaries combine the present of the past and of memories, and the pure present; today's anniversary, that special, auratic state that radiates at the right moments. I have had the chance to observe, and I expect that it is a generalized experience, that birthdays have a certain auratic component. Analysands mention theirs and anticipate it; on that day, they appear to be in a special state that is hard to convey and that permeates the analytic relationship. They impose their pure foreignness as others.

The right time is, then, the time that creates an experience that is unlikely to be repeated. It could be the beginning of a relationship (point of departure), a couple's infatuation, a moment that creates a bond that all of a sudden changes a person's life. It is marked by comments such as, "the day she told me", "the day he looked at me", or "what my analyst said to me". I am reminded here of Marie Langer, who could bring about such experiences. Many people remember a phrase uttered during an encounter with her. (Marie Langer was a Viennese psychoanalyst who fled Europe and immigrated to Argentina, where she became a founding member of the Argentine Psychoanalytic Association. In 1976, she went into exile in Mexico because of her political views. She remained a very active psychoanalyst and very involved in social issues, both in Mexico and in Nicaragua. She left her mark in Argentine psychoanalysis.)

Can we explain the power of the right moment with transference (repetition) theory, or does it belong to a different dimension?

In the context of my previous discussion, it would seem that this theory does not encompass the phenomenon of the encounter of two othernesses. The right moment has to do with an epiphany, while a process of understanding requires slow and systematic reflection on a situation.

The right moment and experience

The encounter, the right moment, is registered as experience. For Benjamin, who takes distance from Kantian analysis and agrees with Agamben, since the First World War

> experience has fallen in value ... Was it not noticeable at the end of the war that men returned from the battlefield grown silent—not richer, but poorer in communicable experience? ... For never has experience been contradicted more thoroughly than strategic experience by tactical warfare, economic experience by inflation, bodily experience by mechanical warfare, moral experience by those in power. [Agamben, 1993, pp. 7–8]

That war, moreover, was followed by others, no less depriving of experiential qualities.

After an extreme experience, certain aspects of what may be shared and communicated are lost. In this case, the potential of the experience to become knowledge vanishes, as does its communicability. Furthermore, it loses the ability to generate an aura. We might, therefore, associate experience and aura, and consider each experience as a unique, new, creative, and authentic production. It is likely that nowadays the improved technical reproductive ability will take a larger role, but there will always be a difference between authentic experiences and works of art, and their reproductions (Benjamin, 1969). It is also true that we need both new productions and reproductions. The advantage of the latter is that many more people have access to them, which leads Benjamin to say that they bear a social and perhaps a political value (1969). Reproductions are also valuable in the realm of learning; they facilitate students' access to original texts.

Yet, we should add that, according to Agamben, the experience of manufacturing work and the role the latter plays in present-day society have introduced a "modern concept of time . . . a secularization of rectilinear, irreversible Christian time" (Agamben, 1993, p. 96). We are dealing here with a process structured according to a time of specific instants, that is, *aion,* the pure present, devoid of previous inscription. I have drawn from Agamben's ideas outlined in *Infancy and History* and Deleuze's *Logic of Sense* (1990) to address the question of temporality in its various aspects. These authors thoroughly review the changing concept of time across history. They posit the difference between circular time and a time that takes place in the present and constantly creates new futures and pasts. The concept of *aion* refers to this time and creates two simultaneous meanings. There is an opposition between *chronos* and *aion* in so far as they refer to two different contexts: the context of narration and the context of events. This approach endorses the idea of the generation of new marks, such as the one produced by the role of manufacturing work in contemporary society. The advent of manufacturing work constitutes an event that breaks with prior history and introduces new focal points, new rules, and new conditions to belong.

Context of discovery, context of justification

A work of art can be reproduced, and the right phrase may be uttered again, yet this will not happen in a context of discovery, but of justification. I have assimilated the idea of the authentic work of art with that of the authentic experience, and the idea of aura with that of the right time, of aionian temporality. As I already mentioned, the right time bears an aura. It construes an experience, and must be somewhat communicable, even though not in its full semantic scope.

Some events are inscribed in a temporality that belongs to the context of discovery. They open up a new world. Others enable us to work thoroughly towards understanding a certain issue. This is what we do in analysis when we tackle the past that is accessible through the present. These two contexts of subjective production are comparable to what Kuhn (1962) designated as "paradigm break" and "normal science". One trend disrupts continuity; the other consolidates the given. An analysand might say, "Even though you had already told me this many times, only now do I realize that . . .". Or sometimes, an even worse blow to our narcissism, an analysand tells us that a conversation with a friend enlightened him, and when he relates what his friend told him, we notice that it is very similar to what we have told him repeatedly . . . Then, what happened? Probably the way he listened to his friend was very different, and the revealing phrase was uttered in a context capable of producing an effect. What is relevant here is "the right time".

Perhaps we utter some very sophisticated remarks while failing to identify the right time, a permeable instance. We do not take into account that a relationship is established in a specific context where giving and receiving are made possible. Outside this context, such remarks could appear as tactless.

Reproducing and repeating

Are reproducing and repeating the same thing? In a certain sense, they bear similarities. None the less, within our framework, repetition is a useful signifier that betrays a conflict, while reproducing denotes the impact of the new and the desire to reappropriate it. Reproducing is copying with a personal hallmark, and perhaps it requires that the person who copies is capable of identifying with the original, performing a task that demands, if it is a work of art, a strong connection with it. Yet, this is no longer the original encounter with the creative experience. In learning processes, copying is sometimes associated with difficulties, but it differs from the incorporation of something new, which demands a different kind of work.

Two possible procedures are activated here. One is closer to what nowadays we call "cut and paste". Through the other, reading an original text gives rise to a new text, a new way of thinking of it, or even the ability to pose new questions based on the discovery of issues the text did not take into account. In our discipline, this process happens often in the reading of Freud's work by those thinkers who are seen as authentic creators. Rereading Freud to discover what he did not say, or to be dazzled by what he said, or even to have him say what he did not develop theoretically, is very different.

Kairos--*to conclude*

The time of the right moment, the time of decisions, often arbitrary, the time of affects, unpredictable and impossible to define precisely, belongs to the temporality of *kairos*. It acquires a name when certain events must be grasped. Then the meanings suggested by the new situation prevail. Appropriate strategies must be identified to do something with what has happened, and these strategies depend on the place and time occupied by the subjects involved.

Perhaps in the case of psychoanalytic practice, *kairos* is the time of our interventions; the time of decisions often untranslatable because they occur within the analyst–patient link, in the subjective constitution of pure becoming. It is also the time that we try to rationalize resorting to our theoretical background when we explain why we intervened at a particular moment.

Kairos is the time that bears a precise meaning for each of us. It is the time of dates, of anniversaries, of specific moments in our lives. Subjects and groups inscribe a kairosian time of their own that is unique to each situation and to each link.

Note

1. Translator's note: the Spanish word "acontecimental" is the translation of the French "évènementiel", a term employed by contemporary French philosophers (Badiou among others). The most widely used English translation is "evental".
2. Translator's note: in Spanish the word for "recovered" is *reencontrada*, literally, "re-encountered".

References

Agamben, G. (1993). *Infancy and History: The Destruction of Experience*. L. Heron (Trans.). New York: Verso.

Agamben, G. (2000). *Means Without End: Notes on Politics*. V. Binetti & C. Casarino (Trans.). Minneapolis, MN: University of Minnesota Press.

Arendt, H. (1958). *The Human Condition*. Chicago, IL: University of Chicago Press.

Badiou, I. (1999). La scène du deux [The scene of two]. In: *De l'amour* [On love] (pp. 177–199). Direction Ecole de la Cause freudienne. Paris: Champs Flammarion.

Badiou, I. (2003). *Saint Paul: The Foundation of Universalism (Cultural Memory in the Present)*, R. Brassier (Trans.). Palo Alto, CA: Stanford University Press.

Benjamin, W. (1969). The work of art in the age of mechanical reproduction. In: *Illuminations* (pp. 217–251). New York: Shocken Books.

Berenstein, I. (2001). *El sujeto y el otro, de la ausencia a la presencia* (The subject and the other, from absence to presence). Buenos Aires: Paidós.

Berenstein, I. (2004). *Devenir otro con otros(s) ajenidad, presencia, interferencia* (Becoming an/other with an/other(s). Foreignness, presence, interference). Buenos Aires: Paidós.

Deleuze, G. (1990). *Logic of Sense (Continuum Impacts)*, M. Lester with C. Stivale (Trans.). New York: Columbia University Press.

Espósito, R. (1998). *Communitas. Origen y destino de la comunidad* (Communitas: origin and destiny of communities). Amorrortu Editores, 2003.

Gonzalez, G. (1996). Cado uno por su lado. Album: *Tranquilo* (singer, Frankie Ruiz).

Isaacs, S. (1948). The nature and function of phantasy. *International Journal of Psychoanalysis, 29*: 73–97.

Kuhn, T. (1962). *The Structure of Scientific Revolutions*. Chicago, IL: University of Chicago Press.

Laclau, E. (2008). Debates y combates. Por un nuevo horizonte de la política [Debates and combats: for a new horizon of politics]. Buenos Aires: Fondo de Cultura económica.

Lévi-Strauss, C. (1966). *The Elementary Structures of Kinship*, R. Needham (Ed.), J. Harle Bell, R. Needham & J. Richard von Sturmer (Trans.). Boston: Beacon Press, 1969.

Lewkowicz, I. (2004). Clase sobre Acontecimiento dictada en el Departamento de Familia y Pareja de ApdeBA (Class on "The Event" taught at the Department of Family and Couples, ApdeBA). Unpublished.

Lores Arnaiz, M., & Puget, J. (1981). El principio de continuidad genética: algunos problemas epistemológicos (The principle of genetic continuity: some epistemological problems). 1ras. Jornadas Argentinas de Epistemología del Psicoanálisis (First meeting of the Argentine Association of Epistemology) *Actas ADEP*: 13.

Nancy, J.-L. (1986). *L'oubli de la Philosophie* [Forgetting philosophy]. Paris: Ed. Galilée.

Puget, J. (1994). De qué infancia se trata? (What childhood are we talking about?) In: *Temporalidad–causalidad–determinismo. Lo reversible y lo irreversible* (Temporality–causality–determinism. The reversible and the irreversible) (pp. 247–273). Buenos Aires: Paidos.

Puget, J. (2001). Nuevas dificultades: lo idéntico y lo múltiple (New difficulties: the identical and the multiple). *Revista de la Sociedad Argentina de Psicoanálisis, 4*: 115–124.

Puget, J. (2002). Qué difícil es pensar. Incertidumbre y perplejidad [How hard it is to think: uncertainty and perplexity]. *Psicoanálisis ApdeBA. Dolor Social* (Social pain): 129–146.

Puget, J. (2003). Intersubjetividad. Crisis de la representación [Intersubjectivity: crisis of representation]. *Psicoanálisis APdeBA. XXV*(1): 175–189.

Puget, J. (2006a). The use of the past and the present in the clinical setting: pasts and presents. *International Journal of Psychoanalysis, 87*: 1691–1707.

Puget, J. (2006b). El presente de la historia, la historia del presente [The present of history, the history of the present]. In: L. Glocer Fiorini (Ed.), *Tiempo, historia y estructura. Su impacto en el psicoanálisis contemporáneo* [Time, history, and structure: their impact on contemporary psychoanalysis]. Buenos Aires: Lugar Editorial/APA Editorial.

Sabina, J. (1990). Con la frente marchita (With withered brow). Album: *Mentiras Piadosas* (Pious lies). www.joaquinsabina.net

CHAPTER SIX

The impact of the time experience on the psychoanalysis of children and adolescents

Ingeborg Bornholdt

Although time experience is part of psychism from the newborn period to old age, the notions of present, past, and future (as differentiated time dimensions) emerge gradually during personality development. With the purpose of discussing the impact of time experience on the psychoanalysis of children and adolescents, I focused on two convergent lines of investigation: the construction of temporality during development and some possible impacts of the identifications with the current adult world.

As early as 1927, in *The Future of an Illusion,* Freud stated that man is "built up from the material of memories of the helplessness of his own childhood and the childhood of the human race . . ." (1927c, p. 18). "What he is entering into is the heritage of many generations, and he takes it over as he does the multiplication table . . ." (*ibid.,* p. 21). The current studies on transgenerationality embrace this idea of unconscious transfer of stories, knowledge, goals, and anxieties that belonged to parents and to the chain of previous generations. The gradual conquest of temporality can also be understood through the exchanges between the baby and his/her environment: the mother, according to how Winnicott understands it. The basic experiences with time take place during the interplay of deep and fundamental relationships of psychism. Therefore, I would like to analyse some important aspects of the gradual development of psychism, considering temporality as one of its outcomes.

Birth is the first great separation; it is a rupture. It is a previous experience of continuity in which we can imagine that the unrecorded time is broken up and this rupture causes anxiety.

This original state of anxiety is analysed by all the authors who study development, since it is what instigates it. I would like to recall only some designations, such as *Urangst*, created by Freud (1926c), which linked this original state of anxiety to the separation from the breast; or "annihilation" by Melanie Klein (1946); "ruptures in the sense of continuity" and "environmental impingement" by Winnicott (1960); "nameless dreads" by Bion (1962); "mental state of depersonalization" Meltzer (1975). These mental states of primitive anxiety originate from the separation of the object when the human being still cannot express his experiences in words or thoughts. The psychic pain caused by separation originates from the sorrow for the loss of a previous experience, from the *past*, that is missed in the *present*. The most primitive time experiences connect themselves with the first elaborations of the separation anxieties. The re-experience of these anxieties during psychoanalysis is remarkable. Many clinical reports start as follows: "after the weekend separation . . .; after the first vacation period . . . after your trip . . .". Meltzer (1967) wrote *The Psychoanalytical Process*, in which it is possible to follow the recapitulations of the previous development stages and the time experience within each analytical process, regardless of the patient's age.

From the newborn to the baby, from the baby to the young child, and from the child to the adolescent, it is possible to observe the gradual development of the psychism, or the ego, as well as the time experiences. Based on certain environmental signs (noises caused by the mother's movement while preparing to breastfeed, for instance), the baby reveals surprising abilities to identify such signs, attribute a reassuring or anxious meaning to them, and react to these signs. The maternal *presence* triggers different reactions in the child. Therefore, in the very early phases of development, the baby is already able to correlate something from the *present* with something from the *past* that was included in his records.

The vulnerable state of the human being is not only the force that impels the psyche to higher levels of evolution, but is also responsible for placing the individual in a perspective of absolute dependence for existence. Since the mind is still unable to differentiate the self from the object, these primitive stages have deep roots within narcissism. The world has not become something outside the baby yet; the world is an infinite time in which the baby, cared for and in the presence of the mother, can sleep.

However, this timeless immensity can instantly become a place–space–time that swallows the baby: a thread of annihilation. Again, the maternal reverie with the sensorial experience of being sheltered in the mother's arms, the touch of skin, the contact between nipple and mouth, the eye contact, restore the illusion of unity, favouring the elaboration of the separation.

The natural helplessness and fragility find comfort in the illusion of fusion, in which there is an experience of completeness. The desire for union (representing a more continuous previous state) follows the psyche throughout life.

The self and the environment, harmoniously mixed, form a primitive relationship of an interpenetrating nature (Balint, 1968). After birth, there are objects with contours and limits towards which the libido (that previously flowed freely) concentrates and becomes rarefied. The environmental faults are expressed in terms of object relationship.

> The origin of the basic fault may be traced back to a considerable discrepancy in the early formative stages of the individual between his bio-psychological needs and the material and psychological care, attention, and affection available during the relevant times. [*ibid.*, p. 22]

Gradually, the mental representation of the object emerges. In the beginning, the object is felt as part of the subject, later on, it is partially differentiated, and, finally, the differences between the self and the object are defined. It is a journey of psychological development that reaches its highest point in a state of relationship with the total object. At this stage, there is also the consolidation of temporality as a result of the elaboration of the mourning due to the "interpenetrating mix-up".

The following report is an example of this state. Six-year-old Peter is a boy submerged in his narcissistic world who was barely able to consider his analyst as a human being. However, he vented his frustrations on his analyst and used him as a type of regulator of his own wellbeing. During more than one year of psychoanalysis, he would not greet his analyst when he arrived and before leaving. It seemed that he did not record the separations, although many times he reacted to them in a rather intense manner, showing psychosomatic reactions and mental states of devitalization after he came back from the first vacation period, for example. Then he would just curl up in the foetal position, asking, "How long is it gonna take?" to finish the session. However, this repeated question indicated a certain record of linear time. The presence of the analyst's mind receiving these meanings without invading the patient with too much interpretation about them cheered up the patient. The questions were disruptions inside the continuity experienced during his sessions and they were present in order to be gradually elaborated. During the second and third year of psychoanalysis, Peter improved his ability to experience goodbyes. He knew when the session was finishing by saying that his analyst "looked at him trying to say that it was time to put things away". He started building small "detectors", such as leaving the soap at a specific position in order to check on the next day whether it had been used. The time experience with external presence and absence began to be differentiated.

The gradual elaboration of the separation anxiety allows for a slow construction of the child's subjectivity, which includes temporality. The unconscious logic (of which timelessness is a characteristic) prevails in the indirect proportion of the child's age. The human being's increasing ability to rebuild and restore the lost objects, or, in other words, the hope in the reconstruction of a time of *presence* of the integrated object, broadens the psychic frontiers. On the other hand, the irreparable objects, as Baranger (1971) describes them, attract the whole temporality to the *past*, making it difficult to achieve the temporal notion of future.

When treatment develops, the internal presence becomes solidified, which improves the child's and/or adolescent's subjective ability. This sophisticated representation can only gradually and partially emerge during childhood and adolescence. We feel intensely the need of dependence bonds.

Let us look back at Peter, who remained in a psychic world where he imagined himself again in the continuous presence of an omnipresent maternal object that was himself in his omnipotent and arrogant self-important state regarding those who surrounded him. At the age of seven, his capacity to acknowledge is slightly improving.

This might happen due to a basic confidence and a solid relationship of a mind able to contain another mind ♀♂ (Bion). When the experience of absence is already able to connect to another experience of (internal) *presence*, it is possible to build better abilities of anticipation of the resurgence. Then the symbolic abilities enhance temporality in the three dimensions.

Temporality and structuring of the depressive position evolve together, and one regulates the other. According to Meltzer, Bremner, Hoxter, Weddell, and Wittenberg (1979), the baby tolerates the separations and allows "the parents to leave to produce another baby" when he/she is able to replace the projective identifications for the introjective identifications. With the help of the introjective identifications, the baby endures and tolerates the separation because he/she can count on a good object that is firmly established inside himself; therefore, being able to identify with some aspects of this object. The ego builds up concomitantly to the initial notions of the *present* (which is based on the memory from the *past*) and it develops the ability to *wait* for the reappearance of the object in the future. The *future* emerges as a possibility of representing and *waiting* (instead of despairing), of repairing and affectively finding again the same emotional state of the contact between nipple and mouth. The concept of *future* acquires the meaning of restoring the union, therefore becoming the integrating matrix.

The child's own abilities articulated to the mother's reverie (or also the analyst's reverie in his job) build up their own continent in the child's mind in order to shelter objects which present more integrated characteristics. With environmental support, the fragile mental states can develop gradually. The time experiences re-edit during treatment, in the setting that allows for the development of transference–countertransference. Continuity, frequency, and a space containing the presence of a mind that can shelter the other mind in a climate of intimacy and, at the same time, neutrality and limit, so that decoding the emotional states generated is possible, represent a space where the time experiences can be lived.

There is something (the mental state of union and continence) in the *past* that is sought to be met; that is, that needs to be created one more time. The growing up and life vicissitudes will require that this state be re-established, restored, reinvented, symbolized in another time and with other elements.

Once more, we return to Peter, who, at the age of eight, presents better possibilities of experiencing alterity. The defences surrender a little and allow for the emergence of feelings originated from the experience of not being the only and omnipotent, but, instead, being human in the widest sense: someone who needs to establish relationships with the others. At the same time that the notion of space emerges, the notion of time also develops. Puzzled, and even fascinated at this point, he realizes: "Can you imagine a tiny little thing such as a cell inside a microbe? . . . what about the *size* to the moon? And to the universe?" Together with these discoveries come the temporal notions: "If I say *now* it has already gone . . . let alone the *infinite*!" He gets anxious: "No, it's impossible to imagine that." By this time, he closely observes the analyst and notices the minimal changes, such as haircut, eyebrows, which shirt the analyst had worn before with these trousers, and so on. He asks the analyst many questions about his age, because now he realizes that the analyst is probably older than his parents. Again, there are waves of anxiety regarding the contact with aspects of reality: "But you are not old enough to die", and gets frightened by this new possibility. The psychic apparatus starts functioning more and more at the level of secondary process, which is less regulated by omnipotence.

Peter illustrates the paths that the construction of temporality follows in his development. The rhythms of the baby's own body (such as cardiac or digestive systems), the rhythms of the maternal body and those related to the care received (the setting when we are in the analytical field) start gaining meaning and effectively represent continuities. The temporal guidelines are built based on these rhythms of interaction between the pairs. Regular rhythms provide safety and continence. As will be mentioned next, currently very little importance has been given to the continuous and regular rhythms.

The notion of *present* and the construction of thought occur based on the elaboration of the lack or absence of something internal, something from the past. Freud (1916–1917) teaches that, with the introduction of the principle of reality, the cognitive process makes it possible to bear the increase in the tension. The thought *anti*cipates something by *post*poning the immediate satisfaction. Before language acquisition, the states of anxiety and primitive sensations are printed in the ego as "representations". The helplessness due to the non-union and the "non-breast" becomes a thought, according to Bion (1962). The thought emerges as a way of making frustration bearable; it tries to restore the absence and, therefore, it is also a process of temporality construction.

The gradual time experience builds up short lines–distances–times that connect something from the *present* with something from the *past*. Minimal distances of time and space between the self and the object constitute a type of channel, or path, where the psyche travels throughout life, meeting the objects and establishing relationships with them.

The triangulation, or the Oedipal scenario, is crucial for acquiring the ability to perform secondary processes. The time experiences associated with the ability to wait are minimal in the young child and evolve together with the child.

The father represents the limit, the reality, and also the temporality. Since Freud wrote *Totem and Taboo* (1912–1913), this function has been widely studied in psychoanalysis. The limit makes it possible for the mind to enter subjective time. The ability to recognize *another*, with respect to mother, is a result of the differentiation between the self and the object and, in turn, it also means the differentiation from another gender and the whole imaginative life of the relationship between mother and father. The elaboration of the Oedipal conflict involves renunciation, postponement, and adjournment. Something from *now* needs to be kept away and it should wait until *later*, until the *future*. This elaboration involves issues such as, "I'm not everything", "there are others", "there are differences"—between genders and generations. Just then the child becomes able to differentiate more clearly among the temporal dimensions. Before that, it is common for the child to say "yesterday" or "tomorrow" when he/she is only able to discern that today's session will finish and will continue later. "Yesterday" or "tomorrow" are only differentiated from the current moment, and can represent the next week or the previous year. Words only indicate the record of a time experience with different dimensions in the internal world.

If the subject is able to elaborate the (Oedipal) castration anxieties, he/she achieves identification with the values of the parental objects and can proceed to the secondary objects.

The omnipotent and narcissistic fantasies of the ideal ego are gradually replaced by the ego ideal and the search for knowledge. As a result of the identification with the parental figures, the ego ideal establishes the ability to try to achieve goals, purposes, objectives, and aspirations (Hanly, 1983). Destiny becomes something to be achieved, searched, conquered: a *future goal.* The child's narcissistic system and the perfections inherent to the infantile omnipotence always need to be further elaborated.

Care and love with respect to the object also constitute strong forces for the Oedipal renunciation and the necessary repression. The principle of reality is gradually established. This originates a slow development of the abilities to establish asymmetric relationships and broaden the consciousness.

The interdiction and the Oedipal elaboration indicate the beginning of the exogamous period (Urribarri, 1998) of latency, during which the sublimatory forces develop the ego. In psychoanalysis, the time experience increasingly assumes the condition of the real.

In terms of evolution, puberty is the beginning of a new period of turbulences that disrupt the previous adjustments. The sublimated infantile sexual drives emerge vigorously at the same time as the genital condition and the internalized prohibitions develop. A set of internal and external reasons disorganizes the time experience again during the adolescent crisis. Freud (1905d) thought this long development happened in two periods of the object-choice. The normative processes continue to occur and their solution can only be found with the elaboration of the conflicts of adolescence (Blos, 1991).

The painful processes of deidealization reach their highest point during adolescence. Kancyper (1990) considers that the identification with the primary objects and the previous ideal objects arrest the psyche inside a "forever" (timeless). The deidentification and the resignification during adolescence require the abandonment of the idealized images of parental objects, which allows for new identifications. During this long period of narcissistic negotiations of great battles and threats (affective, verbal, and material), there is an ambiguous temporality (*ibid.*). The elaboration of the adolescent crisis builds up and installs a more structured and integrated temporality as a consequence of an extensive temporal collapse. The adolescent disorganizes himself/herself in order to get reorganized based on the resignifications that he/she are able to perform. In the psychoanalytical field, the patient can fully experience the temporal chaos. If the psychic turbulence can be endured and contained, new abilities emerge for the adult's productive life.

The space–time frontiers, with their blurred boundaries in the initial periods, become more accurate and differentiated. Therefore, the logic of consciousness gets stronger.

However, since psyche is a bi-logical structure (Matte-Blanco, 1975), it concomitantly involves timelessness and temporality. The ability to consciously recognize the asymmetries (therefore, the possibility of differentiating *past, present,* and *future*) develops at the same time as the symmetric relationships are established. The asymmetric relationships become a continuous process of development during the individual's childhood, adolescence, and adulthood. Thus, an object has *simultaneously* current and real characteristics as well as past characteristics. In the internal world, there is a continuous movement from reality to regression and from regression back to the objective.

This long visit to the movements of temporality construction in the child and adolescent's normal development aims at illustrating the time experiences in the psychoanalysis of children and adolescents, where the body memories and the different stages are widely re-experienced.

The more we allow ourselves to slip into the world of unconscious phantasies of the pair, the more the mind tells us what happens among them (Ferro, 1995). Thus, we can reach redoubts of primitive time memories and give them more elaboration opportunities.

The neurotic circularity tends to be gradually transformed into linearity, which progresses from something from *just now* to something from *also later.* However, the connections are less linear and present forward and backward movements, although an increasing progression occurs when there is positive evolution or an "expanding universe" (Bion). When the child is already able to differentiate between the self and the object, inside and outside, past and present, he can often go back to previous stages where there is simultaneity and symmetry.

Another fragment of Peter's treatment can illustrate the simultaneity of symmetric and asymmetric concepts. At the age of ten, the patient is at a stage of analysis during which the central topic is about a real and significant loss he is currently facing. His emotional state is particularly quiet and depressed, and he starts counting the number of heartbeats on his pulse. He makes a comment: "I think that I have x beats per minute." However, he becomes worried and, then, pointing with his chin to the analyst's pulse, declares, "Check if I got it right." This is an experience of symmetrification and lack of differentiation of bodies and identities between Peter and his analyst, who could thus count on his own pulse Peter's heartbeats. These experiences were taking place at the same time as solid space–time differentiations were being achieved.

Another child, a younger girl who presented a good and confident mood, used to pretend she had a universe of wonderful things, such as a bakery, a toy store, a day care centre with many babies. Therefore, the analyst was supposed to visit her to get to know and be surprised at all those things. Suddenly, the patient walks to the door and leaves the room: "Wait for me, I'll go poop." She immediately came back and said, "The poop did not finish coming down ..." She kept on playing, but she was restless and uncomfortable, stepping on one foot and then on the other. Then, suddenly, she commented, "Tell me when the poop finishes coming down ..." Again, we can notice a regression there. There is a fantasy that the analyst could feel in his own body the patient's bodily sensations. At the same time, however, the young patient was moving in a world where she was already an admirable grown-up who could offer an internal world that generates and feeds.

There is a development, a continuum in the individual's mind, that is similar to humankind's memory. The layers that are closer to consciousness get increasingly linked to those deeper "underground" layers that are far away from consciousness and are located in the baby's first emotional experiences.

New resignifications make it possible for the adult and adolescent's representations to reach, through the layers, the more and more distant childhood. The continuous elaboration of conflicts throughout the whole life allows for repeated achievements of new forms that Bion (1965) named "transformations".

The ability to recreate the object (achieved in the depressive position) makes it possible to acknowledge the loss of the object in the objective reality and allows for the survival of the ego that is able to deal with this reality. This ability provides better conditions for realization.

The object constancy and the contact with objective reality are closely related, and they make it possible for the object's faults to echo, and find comfort in, internal and subjective "presences". The experiences are stored in the unconscious, and from there they have an impact on the current situations.

It reminds me of children and adolescents whose different symptoms could be linked to the elaborations of separation anxieties.

Paul, five years old, had good intellectual development, was extremely serious considering his age, and started to present increasing difficulties in separating from his parents. With great suffering, he remained quiet and curled up when his parents left. He would remain distressed and would wait for his parents, and occasionally he would have fits of anger. He is demanding and accusatory regarding his parents, establishing psychic scenarios where mutual narcissistic demands, guilt, and resentment predominate. Then, Paul developed the following symptom: when he had to deal with separations he became pale, had dark rings around his eyes, and would pick up tiny things from the floor as if he were disconnected from objective reality. During his sessions, this anxiety is transferentially acted out. When the session was almost finishing, as if he were in another world, Paul would suddenly kneel down and start looking for something on the floor in silent despair. He would get anxious, and would hold threads of carpet or pieces of chalk he found on the floor as if they were something valuable and vital for him. There is the fantasy of definitely losing the object. The temporal dimension of *future* still is not integrated and the separations mean something like "never again", or death. The small pieces of rubbish represented his despairing attempt to keep the object.

A seventeen-year-old adolescent says he feels as if he were "swimming in a swimming pool full of glue". With this metaphor, he illustrates his internal world, where he is "glued" to the primary and primitive objects. He has difficulties in creating new representations and symbols through which the time experiences could flow, instead of melancholically retaining him in an illusory past time.

Nathalie, an intelligent and gracious adolescent patient, had bulimia, and started her psychoanalysis in a very fragile physical and emotional state. She was cruel in terms of the transference, due to the level of destructiveness she showed towards herself and the analyst. She swung between fantasies of being mixed up with the object (being a part of it) and omnipotent fantasies of non-existence of the object and, therefore, she could barely consider it. It was as if she were, on the one hand, in a constant state of attack against herself and/or the objects and, on the other hand, as if she were in the deepest state of fusion. There were periods of her treatment or fractions of time during the session in which she would systematically disagree with the analyst's interventions, being very arrogant. She establishes symmetrical relationships. According to her point of view, the analyst needed to be a part of herself: she should not have her own words and her own life. She managed this relationship in an omnipotent manner, transforming what she agreed with into something that she had said.

In her fantasy, she took herself to distant times: she was either in a world where she could act and demand as if she were a little girl, or she talked about grandiose *future*. It was a narcissistic *future*, where she could see herself connected to people as wonderful as herself. In fact, these are representations of her idealized internal objects. Since they are narcissistic objects of her ideal ego, they can easily change from an idealized object to an extremely nasty and persecutory one.

Although Nathalie talked a lot about the *future*, there was actually an imaginative life of restoration of the *past*, where the infantile narcissism ruled. There was a continuous attempt to refind the mental states of the ideal ego, of "your majesty, the baby". The real notion of *future*, which originates from the elaboration of the limit and the reduced omnipotence since the weaning period, is strongly defended and, therefore, it has little psychic circulation.

Here again, it is possible to observe a temporality with predominant characteristics of circularity: the present goes back to the past due to the repetition and reoccurrence of regressive experiences. Even between this present and this past there is so little difference that these temporal dimensions seem more like condensations with unconscious characteristics.

When the narcissism is not sufficiently elaborated, it does not allow for the acknowledgement of reality, which increases the omnipotence (that only reflects a high level of impotence/helplessness).

In these situations, patients of all ages can build up defences against dependence. To this difficulty of recognizing the limits corresponds the other difficulty of building notions of temporal, sequential, and *future* order. Often, the restraint and the limit of the analytical setting (which is, above all, a mental state) provide a certain degree of temporal order and space–time differentiation. In Nathalie's grandiose self, for instance, time has the dimension of eternity. Death as the limit of physical resistance, the excessively low weight, the drugs, or the other abuses, are denied. At other moments, the fragility, which is usually split and denied, imposes itself and seems to be the patient's only psychic reality, taking her to experiences of great helplessness or complete dismantlement.

The fantasies about the *future* of patients with narcissistic pathologies are similar to those of children when they pretend to be fairies, heroes, kings, queens, warriors, super athletes, terrible warriors, witches, beautiful models: it is a completely good or completely bad *future*. It is even possible to find children with poor ability to play, who show a great internal void and boredom; something similar to a complete evacuation of objects. The present did not restore the *past*, and it is not even able to integrate aspects of the objects and the self. There is a domain of the absolute. The individual remains in the eternal *present* or in the eternalized *past* with ideal objects. The denied limit and helplessness create fantasies of an exuberant immortality, or even a vacuum of objects. In this psychic state (located in a deeply *past* time, although it is experienced as something current), there are no frontiers of reality and temporal flow. The omnipotence and the absolute (characteristics of the ideal object and the ideal ego) establish reflective and symmetrical relationships. The infinitude characterizes the object as well as the self. In situations like that, the collapse of the imaginary spatial and temporal unit cannot be elaborated (Kancyper, 1997).

During psychoanalysis, children often act out this denial of reality and limit. They easily close their ears when they do not want to (or cannot!) listen to the interpretations or interventions that make them anxious. Sometimes, they actively sing, close their eyes, act out in several manners how much they need to defend themselves from what is happening. These are devices to act out the −K (Bion, 1962). They act out meanings such as, "I will not acknowledge that", "I close my eyes and this way I can't see", "This doesn't make me anxious; instead, I'm singing! . . .", "It doesn't exist . . .".

Temporality is like an underground fountain whose water constantly flows throughout the development and the psychoanalytical process.

Transference revives old conflicts and feelings. The analyst needs help from his unconscious in order to understand the transference, but he also needs his consciousness. He needs to make these resources available to the analytic relationship and to the patient. To interpret the material is the end of this inter- and intrarelationship that is being revived. The patient needs to be free to place his analyst where and when he needs him. The analyst needs to transport himself to this time and place. He needs to return to the *present*, and then he can interpret the material if he considers that this is the timing.

The continuous interpretation of the circles of neurotic repetition can open temporality. The analyst receives the projection of the objects from the patient's imaginary *past*, *present*, and *future*. As the archaic objects can be differentiated from the real objects, a clearer differentiation of temporality takes place in all dimensions (Baranger, 1971).

Current trends and final comments

I started with babies in order to focus on the time experience in psychoanalysis of children and adolescents. Based on the relationship with the primary objects, internal experiences are recorded. With more or less distortion, the experiences with the external objects have an impact on, and model, the internal objects. The relationship with them and between them build what Melanie Klein described as the internal world, our main focus during the psychoanalytical work.

Regardless of the theoretical line we adopt, the psychic scenario of the child and adolescent who are being analysed leads us to the records of their past experiences. The more rigid and fixed the forms, the more the child moves in an internal world with past objects that are currently present in the psychic scenario and in his life. In other words, the more regressive and dependent our little patient is (up to an extreme manner of defence: autistic characteristics), the more he keeps a relationship with objects of little resignification. The internal world that reveals itself in the space–time of each part of the session, in the whole session, or in each longer temporal unit of the psychoanalysis (weeks, months, years), is a world of transferred fantasies and imaginations.

While working with transference–countertransference, we are working in past experiences. What takes place now, in the present, triggers certain parts of the patient's past and, in its turn, touches certain aspects of the analyst's past. The analyst, in his turn, after elaborating on himself sufficiently, can provide it with meanings and understandings with the help of the present and, thus, help to update the emotional experiences in the relationship. The issue becomes increasingly complex, since the child and the adolescent are equally dependent. In the indirect relationship of their age, the patient and the analyst need the real person represented by the mother and the father. The parents are the ones who make it possible for the psychoanalysis to take place when they try to help their child, when they allow for the difficult and concrete organization of several sessions per week, when they arrange for the payment and, finally, when they face the deep emotional demand of offering their child to another adult so that this person can help them regarding what is causing psychic pain. Parents are directly inside psychoanalysis. Especially with respect to young children, they are increasingly included in the psychoanalytical work. Their temporalities also mark the child's subjective construction of time in an intense manner. Different levels of integration of the internal object correspond, due to the identification with them, to different levels of association or disassociation with the *past* (neurosis).

Further considerations regarding the complexity of this issue exceed the scope of this work, but it is important to remember that the network of object relations within the analytical field encloses those of the patient, his parents and ancestors, as well as those of the analyst. From the continuing elaborations of conflicts emerge perspectives of care for the self and for the object. Thus, repairing forces can be used to consider the other, for the continuity of relations, for the care, and for the capacity to enjoy life.

The discontinuities and ruptures of the child and the adolescent's references (including the external ones, such as environmental changes, separations, deaths, etc.), can not only lead to growth when well elaborated, but also present further requests for the forming personality; therefore, all human efforts try to establish—or, probably, to re-establish—a developing harmony with one's environment, to be able to love in peace (Balint, 1968). The discontinuities destabilize the harmonious and involving environment. Currently, we can observe a certain recklessness regarding separation experiences. Parental functions are easily "outsourced", exposing children to successive bonding changes.

I believe that, elapsing from normal development (and during psychoanalytic treatment), the child, the adolescent, and the young adult's notions of *future* are necessarily conceived in a natural and healthy narcissistic perspective. The *future,* during these phases of development, had the magnitude of the subject itself. Only gradually can the temporal notion of *future,* articulated with the mortality of the subject, be elaborated.

Peter, confronting the idea of the analyst being older and closer to death (clearly considering here the transference and the proximity of the end of his analysis) speaks about a scene in a movie that shocked him: he saw a child's tomb. "But fortunately that almost always does not happen . . . ." He calculates the difference in decades between his age and the analyst's supposed age. To him, this distance and interval are huge, which is reassuring.

The memory of the lived experiences flows underneath, and the time experiences follow, the entire psychoanalytic process.

The elaboration of the *non-fusion* and of the weaning is revived and represented again in the Oedipal interdict meaning that *you cannot.* During adolescence, the interdiction and the adjournment acquire the meaning of *not here,* and, for the elderly they mean *no longer.* Throughout life, continuing re-editions of previous anxieties within new contexts originate resignifications. They are continuous processes of elaboration, for which Bion used the model of a spiral in constant ascending and expansive movement. Throughout these psychic processes, the object constancy is the most solid anchorage for each new request and vicissitude. This generates the internal capacity of searching for aesthetic and ethical goals; of consideration regarding the objects and of renunciation of immediacy and consumership (so evident currently).

The current trivialization of violence and sexuality, as well as the trend towards the establishment of symmetrical relations (insufficient differentiations between generations, genders, ethics of what is fair–unfair, what is right–wrong, etc.) generates increasing appeals to states of excitement, to the immediate; furthermore, to the intolerance regarding the construction of the thought through conversation with the other, through reading, and through gradual elaboration. Where everything is possible and allowed, the flow of impressions exceeds the capacity of elaboration. We are experiencing a symmetrification trend involving generations (with increasing attempts to transform children into small adults and adults into adolescents). The confrontation that is important during adolescence for the establishment of non-narcissistic and non-incestuous relationships is being avoided. As adults, we are precariously supporting our children and adolescents in their necessity to be dependent, constructing their subjectivity. Therefore, if the adult is avoiding the processes of elaboration of mourning, avoiding psychic pain and the necessary time for the construction of subjectivity, how will a child be able to do it? The relaxation of differences and limits by the adult world makes the adolescent's reorganization difficult. The general symmetrifying trend leads to the risk of collapse or short circuit in the natural conductors of temporality (*past→present→future*).

Once the structuring of the personality happens, fundamentally through the identification with unconscious aspects of external objects, the increasing tendency of adults to act as adolescents (privileging acting instead of reflecting, haste instead of slowing down, etc.) has serious impacts on the construction of identity and temporality by children (Bornholdt, 2004).

Projects with unconscious meanings regarding the search for ego ideals with internal objects are being run over by others with the meaning of immediate gratification, of "volatility", of "disposability", characteristics of transitory bonds (Araújo, Iankilevich, Bornholdt, & Campos, 2000). They are projects that reflect desires of immediate gratification, similar to the omnipotent narcissistic forces of the ideal ego. These projects narrow the space to the *future*, given that they strongly invest in the present and "consumerism". The integration of temporality, object constancy, the ability to wait, and tolerance of gradual conquests are impaired. The attempt to speed up directly hinders the construction of the internal world, causing an increasing emptiness instead of the accompanying internal objects.

Too little tolerance of being kept waiting and having to go through long journeys, which are necessary for the elaborations to take place, and the trend of not allowing enough processing of feelings of mourning and loss, make it extremely difficult to build the notions of *future* articulated with mortality. It is possible to recognize in this a perspective of insufficient capacities for the integration and repair of the internal object. Pathologies related to addictions, recklessness concerning physical and moral integrity, the ecological damage caused by pollution, are examples of areas in which we can observe the precariousness of the notion of *future*, and, consequently, the lack of responsibility of the unconscious projects emerging from the ego ideals.

Centrally, the difficulty in tolerating the time taken for the development of a personality, of a phase, or a project, results in conflicts regarding the formation of an identity. The issue of time experience in the psychoanalysis of children and adolescents refers us back to our own identity as much as to the construction of identity of these children and adolescents. Taking into consideration all current demands and pressures, we still need time, serenity, patience, and tolerance for the gradual elaborations of the relationship of the analytical pair.

Only when the renunciation of the attacks from the sources of dependence and the recognition of their existence and value subsists, the child and/or the adolescent will be protected and supported by caring and repairing forces of their own.

The presence of the internal object starts building up in the baby who can trust and wait when supported by the model ♀♂. The maintenance of the object in the internal world results from the ability to elaborate depressive anxieties, generating hope that makes possible the concept of *future*. However, the *younger* the subject, the more he really depends on the external object to survive.

From the elaboration of the weaning and separation anxieties to the Oedipus complex, to latency, and to adolescence, there are successive chains of elaborations through which the objective reality becomes increasingly recognized and, therefore, the temporal notions are further integrated. In this mental journey, knowledge expands and the borders of psychic capacities widen.

In short, the harmony between narcissistic supports, gradual achievements of autonomy, and connections with objective reality integrate the personality more and more, and also the notion of *future*. It becomes a time in which good objects can be projected, increasing the ability to achieve accomplishments.

Object constancy, which is prompted during normal development through conflicts throughout the life cycle, makes possible the interlocution with integrated and good internal objects. Thus, even when the objects are absent from external reality, their relationship with the internal world is not breached and the conceptions of reality and temporality keep expanding. The best inheritance that we can leave the children and/or adolescents whom we analyse is exactly the expansion of relationships regarding integrated internal objects. The responsibility for these internal relations makes more likely the search for the development of the individuals' abilities throughout their lives.

References

Araújo, M., Iankilevich, E., Bornholdt, I., & Campos, M. (2000). *Reflexiones sobre destinos de duelo no elaborado. Participação em plenária.* Uruguay: APU.

Balint, M. (1968). *The Basic Fault.* London: Tavistock.

Baranger, W. (1971). *Posicion y objeto em la obra de Melanie Klein.* Buenos Aires: Ediciones Kargieman.

Bion, W. R. (1962). A theory of thinking. In: E. Spillius (Ed.), *Melanie Klein Today* (Volume 1) (pp. 178–186). London: Routledge, 1988.

Bion, W. R. (1965). *Transformations.* London: Heinemann.

Blos, P. (1991). Masculinidad: la rebeldia contra el padre del esfuerzo adolescente por ser masculino. *Revista de Psicoanalisis,* 1(1): 19–30.

Bornholdt, I. (2004). Construções da temporalidade no desenvolvimento normal. *Revista Latinoamericana de Psicoanálisis,* 6: 221–238.

Ferro, A. (1995). *In the Analyst's Consulting Room*. London: Routledge, 2002.

Freud, S. (1905d). *Three Essays on the Theory of Sexuality*. *S.E., 7*: 125–245 London: Hogarth.

Freud, S. (1912–1913). *Totem and Taboo*. *S.E., 13*: 1–161. London: Hogarth Press.

Freud, S. (1916–1917). *Introductory Lectures on Psycho-analysis*. *S.E., 15–16*. London: Hogarth.

Freud, S. (1926d). *Inhibitions, Symptoms and Anxiety*. *S.E., 20*: 77–174. London: Hogarth.

Freud, S. (1927c). *The Future of an Illusion*. *S.E., 21*: 3–56. London: Hogarth.

Hanly, C. (1983). Ideal del yo y yo ideal. *Revista de Psicoanálisis, 40*(1): 191–203.

Kancyper, L. (1990). Adolescencia y desidentificación. *Revista de Psicoanalisis, 47*(7): 750–760.

Kancyper, L. (1997). El resentimento y la dimensión temporal en el proceso analítico. *Revista de Psicoanálisis, 94*(6): 1301–1324.

Klein, M. (1946). Notes on some schizoid mechanisms. In: *The Writings of Melanie Klein, Volume 3*. London: Hogarth Press, 1975.

Matte-Blanco, I. (1975). *The Unconscious as Infinite Sets*. London: Duckworth.

Meltzer, D. (1967). *The Psychoanalytical Process*. Strath Tay, Perthshire: Clunie Press.

Meltzer, D. (1975). *Sexual States of Mind*. Strath Tay, Perthshire: Clunie Press.

Meltzer, D., Bremner, J., Hoxter, S., Weddell, D., & Wittenberg, I. (1979). *Explorations in Autism: A Psychoanalytic Study*. Strath Tay, Perthshire: Clunie Press.

Urribarri, R. (1998). Descorriendo el velo sobre el trabajo de la latência. *Revista Latino-Americana de Psicanálise, 3*(1): 257–292.

Winnicott, D. (1960). Ego distortion in terms of true and false self. In: *The Maturational Processes and the Facilitating Environment*. New York: International Universities Press, 1965.

CHAPTER SEVEN

Time and the end of analysis

José E. Milmaniene

One of the basic objectives of psychoanalytical treatment is to create the necessary conditions for the patient to be able to accept his own historical time on the horizon of acknowledging his own finiteness. This effect of subjectivity, which supposes the appropriation of his singular historic time, implies his transcending to the same degree his submission to the alienating time of his parents and his stay in the sterile timelessness of narcissism. (The concept of subjectivity alludes to the complex process of the constitution of the subject. It means the transition that goes from the primitive narcissism of the child, who lacks language and is fused symbolically with the phallic figure of his mother, and is established thus in the drive order until his definitive configuration as a subject of the word. This process of entering into the symbolic order requires the symbolic operation of cut-castration, which should be carried out by the symbolic figure of the Father, who represents the Law.)

Analysis brings liberation from a monotonous, lineal, and empty time, lived in only for the insistent repetition of symptomatic acts and access to a time that does not know this sameness. After analysis, the time of egocentric self-reference should give way to a time that is tightly linked and committed to the social bound, and it is there that the subjectivity of the patient is registered.

To analyse oneself is to free oneself from a viscous temporality, marked by narcissistic mono-themes, and to enter into the unfolding of decided acts inside a full time that can be enjoyed in its pleasurable passing, beyond anxious precipitation or the apathy of constant postponement. We can, therefore, sustain that the mechanism of analysis allows the free unfolding transference of time, and on this happening the subject arises, in so much that he is now able to narrate symbolically his own libidinal history without idealizing myths or melancholic relapses.

What is involved is the placing of the patient in the territory of "sublimation", far away from temporal fixations with traumatic events and from remaining with obsessive, inexplicable meditations. To be cured is to place oneself in the time of free decision and responsible acts that acknowledge "the other" and respect his time as much in a liberal and metaphoric sense.

Let us remember, moreover, that the "narcissistic temporality" is bound to the compulsion to repetition, which means that that the subject is folded back over himself, in a frozen time that is at once unmoving and without history. He is incapable of giving himself to, or acknowledging, "the other".

When, during analysis, the acknowledgement of unconscious desire arises, at the same moment, one establishes oneself in the "historic temporality of the Oedipal organization" with regard to the insuperable symbolic distance that is created between the subject and the object of his desire that has been, and always will be, lost.

Thus, we can sustain that psychoanalytical intervention dissolves just as much the imaginary and omnipotent notion of eternity as the notion of timelessness, both of which emanate from narcissism, and it does this through two operations that are inherent to the cure, and makes it possible to place the subject within his own symbolic time. These are:

- the dissolution through interpretation of the figure of the beloved immortal father, an imaginary figure that leaves the subject in a passive and morose wait for an idealized love;
- the horizon of the inevitable end of analysis, which acts as a real and inescapable limit that destroys the believer's hope in

> redemption through his remaining in the transference of love and makes the subject confront the most absolute solitude, which transforms into the confrontation with the absolute master: death.

Thus, the treatment sets in motion an "effect of subject", given that the psychoanalytical discourse makes the being temporal through the "interpretative dissolution" of the symptomatic points of temporal detention, which are correlative, at the same time, with the illusion of timelessness in which the "ego" of the patient is crystallized. On the timeless chaos of "this", a pure flow without a beginning or an after, formed of perpetual instants on a limitless background of unnameable anxiety, there must be made to succeed the sequences of logical times, articulated over a landscaped network of resignations, as Lacan set out (1971, pp. 21–36). (Logical time is the unique time of the subject, and is linked more to chronology, to the conjectural process of the elaboration of vital experience, thanks to the significant order. What is in play is a logical registering in the Freudian resignification–*Nachtraglich*-configured by the instant of seeing, the time of comprehending, and the moment of concluding.)

The psychoanalytical session is, therefore, a privileged space where a genuine relationship with time can be established, a place that is greatly under attack from psychopathological structures, as is seen in their most frequent distortions: manic accelerations, melancholic fixations with a past that is never cancelled out, anxious yearnings, standstills, temporal and indistinct repetitions, simultaneities, and chronological super-positions, etc.

The psychoanalytical mechanism establishes a relationship with time and obliges the acceptance of its symbolic legality, given that the framework, as much as the order that the interpretations impose and the constructions over the confusions and imaginary disorganization, establishes the subjectivity of the patient in respect to paternal law and in the historical horizon of all symbolic legality. That is to say, time acquires its true historic sense, beyond remaining in a timelessness in which one is stuck in symptomatic pleasures.

The retroactive reading that analysis proposes allows the finding of another significance in biographical acts and deeds and includes them in a narration that gives them a new sense in the interior of a temporal order that does not respond to the distorted sequence of neurosis, but rather to a symbolic reordering that alludes to the timelessness of their own projective self-referential interpretations.

Thus, we must insist that the most important element of the cure resides in the relationship that the patient establishes with the "temporal limit", because of the instance that occurs when the "end of analysis" is reached.

The temporal limit—that is always annulled, denied, or contradicted in psychopathological structures—is re-established through the closure that the finalization of the transference relation imposes. The end of analysis makes it possible for time to become historical in act, since the subject is dislodged from any type of infantile dependence on " the Other", based on the omnipotent supposition of the immortality of the subject, just as it was based on the immortality of his parents in his infancy.

The relative negation of the passing of time that tends to install itself in any significant relationship favouring narcissistic tendencies inherent to subjectivity suffers a severe attack with the conclusive severance that is implied by the end of analysis, an act that works as a metaphor for the finiteness that is to come. Therefore, from the moment of this act–event, time acquires its own true symbolic statute, given that consciousness is taken of its unceasing passing, and, thus, the transitory and fleeting nature of existence.

Thus, analysis makes it possible to establish another relationship with time, given that it makes possible the deconstruction of the symptomatic relation with the same, and is able to establish "time as an experience of discourse, that is thus bound to symbolic discontinuities". In this way, the static relationship with time can be overcome, which captures the subject in an alienating net that is not dialectic and allows him to accept the vertigo of a limited, discontinued, and fragmentary time which belongs to a desiderative universe over a symbolic horizon formed by differential oppositions.

To sum up, analysis creates an "empty void" outlined in words and symbols that allude to the temporal limit in which existence is played out, delineated by anticipation, delay, nostalgia, promise, and the nothingness of before and after. Thus, the symbolic registration of the temporal limit for the end of analysis establishes the differential framework between the register of "presence" and "absence", the "phallic" and the "castrated". Therefore, the "not everything" must be accepted in the temporal plane just as the "between" that interrupts the continuous lineal of an existence trapped in the eternal coming of a time without a before or after, which a patient who lives in the acting demonstrates.

The temporal discontinuity that the analyst establishes serves to create an "interval", so as to be able to establish in the same a "silence and a speaking" that recognizes the *jouissance* and the repressed desires that govern existence.

And so, through transferred personal experiences, time can be subjectified in its true socio-historic dimension. This supposes the acknowledgement of one's own past, unimagined and resignified by the symbolic reading of the traumas suffered, beyond depressed complaining and self-referential paranoia and responsibility for the "present", which includes acknowledging one's own desires, a subjective position which grants one at the same time the possibility of confronting, in freedom, the "future".

Definitively, this means man overcoming his submission to quantifying times of social conventions that always have superego mandates, and being able to establish oneself creatively in one's own unique, historical horizon. We must remember, with respect to this, that pathologies marked by an excess in the "drive" order do not manage to establish any kind of temporal movement because the subject revolves in a circle, trapped in his own *jouissance*. The term *jouissance* is used as Lacan used it throughout his work. It means the paradoxical imbrication of the libido with the death drive, which is expressed though "pleasure" and the fascination that is procured through a suffering that is highly eroticized. What is in play is the paradoxical pleasure and unconscious satisfaction that procures any symptomatic or acted dissatisfaction.

The possibility of generating a dialectic movement that circulates from "the alienating timelessness of narcissism" to the "temporalization of time" supposes the resolution in analysis of at least three fundamental dimensions in the life of the subject, and here I follow the developments of the philosopher Emmanuel Levinas:

1. "Sexuality" with regard to acknowledging the otherness in its irreducible difference. Of course, it has to be understood that the experience of time in analysis implies accepting the Other sex with its unique way of confronting its own past. Thus, the "feminine–maternal" conception of time is tightly linked to the influence of the biological body, its cyclic rites, hormones, and pregnancy. But, in contrast, the "masculine–paternal" conception is based much more on historical time and production and is thus more distant from the impositions of the body.

   Anyone who has undergone the experience of analysis will be able to acknowledge and accept intellectually the temporal differences inherent to the different sexes, beyond envy and resentment. The personal experience of time supposes the laying down of the exclusive self-sufficiency of narcissism, so that "time as a pure difference" can be symbolized, which is what is established basically and paradigmatically in the framework of the amorous encounter between a man and a woman.
2. "Paternity" as an expression of a gift to a child is an act that establishes the subject in the succession of the generations. On undertaking the test of paternity, man consummates the absolute destitution of narcissism and accepts in this act his own transitory historical time, as the fleeting "mediator" between his parents and his descendents.

   It is, thus, possible to understand, in both sexes, the incapacity of egocentric personalities to establish a family and take their place in the inter-game of the generations, since being a father supposes the opening up of the temporal space to include in it another being that will survive him and which he must care for and protect unconditionally. Even if a child does suppose a certain recuperation of narcissism, his presence must finally become the ultimate symbol of one's own transitory nature, in such a way that, in this relationship, the tension between the time of the "One" and the "Other" will express itself violently.
3. "Death": an analysis seeks to provoke conditions in which the subject can establish a new existential project and in this way avoid the emergence of melancholy *jouissance* that results in the consciousness of an inexorable end (see the articulate

> considerations of Freud in "Our attitude toward death" (1951, p. 1016). The idea is, then, to accept creatively one's own finiteness, since it is only consciousness of death that obliges one to abandon manic and omnipotent positions that deny any lack and submerge the subject finally in the paralysing illusion of immortality. The depressive acceptance of death as a real limit is a condition for the unfolding of one's own potential, since it obliges one to "think of nothingness". This task is impossible, if it does not arrive in the order of melancholic relapse, and generates different "sublimatory" expressions with respect to creative testimonies for this same impossibility. Creation is nothing more than a defensive strategy to make the "void" tolerable, a "void" that the idea of death causes, doing this paradoxically through words and symbols that replicate—on a spiritual plane with insubstantial materiality of the significants—the real void that we wish to close.

Thus, we insist that analysis makes possible the access to historic time that the subject desires, with the consequent abandonment of the temporality of timelessness of the narcissistic universe. This transition, which is forced from the very timelessness of the egocentric "ego" to the subjectivity in the space of the otherness, generates a violent resistance in the patient, since it not only causes anguish in the encounter with the "other", but it also has an effect on the negative to renounce the masochistic *jouissance* that results in debts that will never be settled, unending guilt, and masochistic and sacrificial resignations inherent in the most regressive states of a being. Therefore, it is not strange that the subject attempts to return to the world of narcissism, full of repetitions without any difference, with the lowest moods of a time that has always "a past of neurosis", a present of acting, or an annulled future of melancholia and passages of suicidal acts.

Psychoanalytic practice creates a privileged space to establish a symbolic relationship with time, purified from imaginary fantasies that either distort or do not know its unceasing passing, since it makes possible the opening of a moment for reflection that enables the emergence of a "time of desire in the libidinal history, beyond the timelessness of narcissism". It is a necessary "interval in time" that opens when one is conscious that life unfolds itself between nostalgia for the past that "could have been" and the utopian promise of a constantly different future, given the structural impossibility of any desiderative realization of the same.

The aspiration of analysis is that the subject will overcome the paradoxes and symptoms linked and binding him to an "imaginary time", and will then be able to enter into the "symbolic" dimension of the same, which includes its own "real" limit represented by the figure of death. When one begins the "end of analysis" one possesses a retroactive, temporal historia-building that establishes the subject at the core of an existence marked out by a creative being that neither dates the past nor idealizes the future, defensive mechanisms built up to flee from the responsibility of a present that calls for committed action in one's desires.

In establishing, during the cure, an "empty space"—an effect of the dissatisfaction of transference demands, and "occupied" only by the insubstantial materiality of the symbols and words—there emerges the "pure nothing that is time itself, with regard to the perception of the unassailable happening on the horizon of one's own finiteness".

In other words, analysis brings about the "metaphorical experience of time", an effect of the insolvable tension between the "timelessness of the primary process with the symbolic temporality inherent in the language of the secondary process".

We can, thus, define the experience of time in analysis in the following way: "only the overcoming of narcissism, thanks to symbolic castration makes possible the transition from a time without time of a egocentric 'ego' to the time of encounter with the desire of the Other, measured by the emergence of language".

By becoming subjective about time during analysis, one acquires the status of an "event", since one stops reverberating monotonously on oneself, anchored to a past that is never cancelled out, to establish oneself in the vertiginous open flow towards the enigma of love and the mystery of death. (I have developed this more deeply in Milmaniene, 2006.)

In analysis, the subject poses the vital question, "To what time does my time want to bind itself?" The answer that the neurotic phantom gives indicates that one always desires to be contemporary with the significant "Others" of infancy. And, despite the fact that symbolic time circulates endlessly, the patient remains stuck and fixed on the Oedipal objects in such a way that the symptoms and his actions impose the anachronism of "the timeless temporality of the unconscious". The temporal paradox lies in the fact that our fantasies live in a mythical past, not only so that we can take pleasure in the benefits of Oedipal dependencies blown up by narcissism, but also to defend ourselves from the uninterrupted passing of time, with the stamps of castration inherent in its passing: losses, ageing, and death.

Psychoanalysis seeks therefore to reinstall the subject in the temporal categories clearly differentiated from the present, past, and future, which find themselves distorted and confused when the subjective scenario is invaded by the arbitrary dimensions of an imaginary time, far from the socio-cultural conventions imposed by discourse.

Psychoanalysis, as a process of sublimation, permits the existential assuming of time and the distancing of oneself from the chaotic temporal whirlwind, a pure void without limit or interval, an abyss without borders, in which unbalanced subjects are shipwrecked by psycho-pathological structures, such as those who suffer extremely the "pathologies of *jouissance*" (addiction, food disorders, perverse behaviours), in which the "excesses without limit of the pure drive order" rule.

I would now like to relate a brief case history in which a patient in analysis could resolve her pathological relationship with time, a severe Oedipal fixation with the death of her venerated father (Milmaniene, 2006, pp. 86–87). She harboured the conviction that she would die when she was a few years older than her father had been when he had died. In this way, she was demonstrating an extreme fidelity to his figure, since she supposes that her own death would have to occur "only" when she was a few years older than he had been when he had died. Perhaps the few years that she would "live more" than her father meant for her "life without him" with respect to the "imaginary" prospect of this time in which she would be able to enjoy with pleasure and liberty the libidinal disinvesting of his figure. This meant an imaginary, limited time during which she would experience the solitude she desired, perhaps not only to recuperate her own independent affections, but also paradoxically to be able to yearn even more for his absent figure. She could only free herself from his image and her fixation with his shadow in this hypothetical, virtual time that would open up into an abyss between the death of her progenitor and the period during which she would live after arriving at this same age; then she would be free of his ghostly, beloved presence, but at the same time still a slave to it.

There were, therefore, three times unfolding in this clinical situation: "real time", the biological death of her father, "symbolic time", "that of the identification" that had become the point of reference for her own age of death, which was effectively that at which her father died, which she would survive for a brief "imaginary time" formed by this pair of years in which she would outlive her father, whose death had become the absolute date that marked the end of her Oedipal ideal, from which point she had begun a phase of depression. The end effect of this incestuous, melancholy position was that the flow of time was stopped, since she had come to a standstill in a "static time without desire".

The work of analysis was to make the patient conscious of the *jouissance* that can be established by fixing oneself in a libidinal history that unfolds itself in relation with a time that is alienated from the pure Oedipal dimension, without her subjectivizing herself through the establishment of her incestuous relationship and with her thus being able to create her own desiderative time, inherent to the construction of her own libidinal history. The task of analysis supposed the resolving of the intensely incestuous link, to make possible the construction of her own temporal history, bound to the enjoyable time of pleasurable encounters with the Other sex. In the course of the cure, a real dissolution of the organization of the biography based on imaginary chronological criteria takes place, and an existence is reconfigured that recuperates symbolic criteria that, elaborated, rescue the truth of the unconscious desires of the neurotic infant history.

To sum up, "after an analysis there can arrive an historical time of pleasure that results in acts and desires that are assumed without guilt, and which are beyond the a-historical temporal fixation with Oedipal *jouissance* of childhood".

It is, therefore, understood that what is in play is the placing of a dialectic in the neurotic conflicts from "imaginary time", formed by tortuous repetitions, rigid libidinal fixations, delays, and retrogressions, the product of the invasion of unconscious phantoms that impose the "outside time" that operates in the "other scenario" unconsciously, so as to be able to begin to push the subject toward the "symbolic time produced by the word" that permits the subject to break up the confusions that are the product of the marked disjunction between the time of those enunciated and the enunciation.

We should, moreover, remember that the deficiency in the capacity to express verbally the representatives of the affects is the origin of the distortion in the temporal axis. This eventuality happens when the fused narcissistic maternal symbiosis is not dissolved by access to a language that allows the operation of a symbolic cutting-castration, that moves paternal law (Kristeva, 1995, p. 108). Additionally, this same situation can be produced when the father eroticizes excessively the link with a child, favouring a breakdown in the exercise of his function, as happened in the case of the patient hereto described.

"The analytical session" offers the patient the possibility of becoming conscious of his conflictive and unresolved relationship with time, the cause and condition of his illness. Thus, it is of the greatest theoretical interest to reflect on the temporal structure of the session in whose framework the rectification and establishing of this dimension will be produced.

It can, therefore, be deduced that "chronological time"—infiltrated by merely imaginary and arbitrary chances—is displaced by the "other" time that is linked to the narrative and biographical conception and is marked by the "symbolic times" which are produced by the conscious establishing of the subject in his libidinal history. This supposes as much a progressive time that links the subject to the "Other" by a complete inversion that brings him back to the "ego" as happens in this pathology. (With regard to this, it is interesting to point out also the hysterical and obsessive manipulations that neurotics elaborate to refer to the anguish that the passing of time causes, as is described by Poisonnier [1999, p. 221].)

In this sense, we can hold that the analytical session should be carried out "between" the stability of a formal, standardized duration in accordance with the style and customs of the work of each person, and the foreseeable duration of each existential encounter between the two subjects is reciprocated in this moment of the development of the treatment. This means articulating in the session the legality of socio-symbolic time with the temporal singularity of each patient and with the proper consideration for the times for elaboration of each person.

The conventional time agreed within the framework makes it possible to put in evidence the different ways of managing time on the part of the patient, and particularly the psychopathic, manic, or melancholic handling of the same.

However, the possibility of "subjectivizing the time for an analytical cure" supposes the condition that the analyst himself has resolved his own relationship with himself, since, if he has not constructed his narcissistic and or Oedipal positions, there will be the temptation for him to conserve the transference link and not to move toward the goal of the "end of analysis".

The indefinite prolongation of analysis, or the carrying out of a treatment without a project with a clear end, serves to deny the undeniable limit of time. We should, therefore, hold that, without a therapeutic policy that includes from its beginning the idea of the end of an analysis, we are at risk of denying the very passing of time in a relationship that, if it is made eternal, will annul the dimension of the lack of time and of mortality, inherent conditions for the assuming of a totally subliminal position.

The temptation to deny the passing of time on the side of the analyst can be consummated through the configuration of a link that it is supposed will never be concluded and, therefore, replicates the figure of functional maternal symbiosis. Let us not forget that, in the interior of the same, there rules an extreme timelessness that has no space for the emergence of the legalizing words that install the symbolic temporal difference and make possible as an end the necessary separation after transference alienation.

What is in play in these cases are narcissistic transference relationships that seek to draw out the passing of time and to put off indefinitely the conclusive act, and they paradoxically produce an effect that is contrary to the one they seek: that is, the assumption of castration with regard to the creative acceptance and the "depressive" of the vital transitory nature and irreversible passing of time.

"The assuming of castration" supposes a complex process that implies the laying down of the omnipotence of narcissism and the a-symptomatic acceptance of one's own finiteness, as it does the symbolization of the sexual difference to the very limit that is imposed by this real residual limit that cannot be elaborated and that Freud called "the living rock of castration": penis envy in women and the anguish of frustration in men.

We should recall, moreover, that the analyst should be conscious of his own existential moment and be attentive to the consequences that his age could cause in the transference link. There are analysts who cannot accept that their own ageing may, to some extent, cause a deterioration or deficit in their capacity to carry out their functions, be it in a sensory or intellectual sense. In these circumstances, they should have the ethical conviction to interrupt the cure and send the patient to another practitioner.

On the other hand, if the analyst does not assume the passing of time and continues to have the attitudes appropriate to anther time of life, he could cause an effect of identification contrary to the one he should be practising.

Psychoanalysis is the great subliminal practice of our time, and its transcendence is based on the fact that it has elevated "loss as an authentic objective possession". The subjectivity of the experience of time in the context of the cure has turned out to be exemplary, since it works as a metaphor for all loss, in as much as it tries to approach an "object" that is essential only when one possesses it as something lost. "To have gone through an analysis is to suppose that one has assumed that what one fears to lose is the loss itself", being conscious of the loss that is the obstacle to the fetishism of time, a fetishism that is inherent to any denial of castration.

To be able to narrate existence analytically, it must be supposed that one has accepted "the lost time" of childhood and that one has acknowledged that the only thing that one can do is to try to put into effect phantasmatically our history through "repetitions with a difference", a truthful discursive recapturing of what has been, now as witnesses of a past which we read not as victims, but as part of a structure that trapped us because of our drive to *jouissance*. To analyse oneself is to construct the metaphor of loss as our only possession, without falling into melancholic relapses or timeless fidelity to the mythic "temporal object" that has been lost.

The experience of time in analysis supposes also the conception of the same as not a mere desexualized abstraction, but as a drive that is temporal, linked to the alternating rhythm of the phallic plenitude and the emptiness of our personal experiences of castration.

Subjective time is founded, then, on the pendulous vacillation that is produced at the instant that lies between the "ego", when it is in phallic expansion, and the contraction of castration, a temporal hiatus in which the after logically precedes the before and gives it another "posterior" meaning, just as Freud continually reiterated throughout his work. The perception of the passing of a subjectivity that fades and is recreated once and again is time itself in its becoming.

We should, therefore, be conscious that our "ego" does not live in a crystallized present, but in the "interval" between the unchangeable before that has passed and the after that has not yet been, and that time accepted existentially implies us thinking without concessions about "the difference between the differences", since the insistent caesura caused by the impassable interval in the irreconcilable dimensions of time, "between the conventional socio-symbolic temporality and the timelessness of the unconscious".

Psychoanalysis has caused a new discursive practice and an unknown style of symbolizing existence, in such a way that a new order has been established, subject to, and settled on, a certain "reordering of time", distant from the "manic acceleration" that is proper to postmodernity. Analysis makes it possible for us to overcome a present without projects and without memory, where subjects who are alienated and exiled from historical time are shipwrecked, and it places us in a measured time which is serene and has a dialogue with an "Other" that is acknowledged and respected in its difference.

Moreover, transcending towards the otherness also allows us to escape from a "temporal standstill and passiveness" that leaves the subject with the illusionary, omnipotent fantasy of remaining in the merged narcissistic unit, previous to time.

A subject who has managed to undergo successively an experience of analysis has formulated with rigour the decisive words that allude to the passing of time and which permit him a symbolic and asymptomatic elaboration, a before/after, yesterday/now/tomorrow. Thus, continuity and the discontinuity of time, its delayed advances and precocious advances in relation to existential chronology, its inaugural totalities or its final fragmentations, its intervals, its inversions, its contractions or drawing outs, its contortions, its accelerations or standstills, are unfolded in the analytical scenario in such a way that the subject can construct a "genuine subliminal experience of time".

"And we can just sustain that to analyse oneself is to assume castration as a subjectivity of time": that is to say, what is in play is to accept creatively the unstoppable passing, without abandoning the construction of a sublimal project that includes love and work, in accordance with the possibilities of each age. By being able to talk about the conflicts and the traumas that assail us, we begin to construct a special–temporal scenario in which we can unfold a subjectivity capable of giving us access to our libidinal objects and which allows us to express our history with coherent narrative fictions that not only consider the past, but also "open" themselves to the future, and make destitute, as their end aim, the present void of unnameable anguish.

The deconstruction of time in which the subject who suffers lives alienated, allows, thanks to the interpretative recourse, the retroactive reconstruction or the libidinal history, purified now from the "excesses" of imaginary senses with which the subject intended plugging up the nuclei of castration, caused by traumatic experiences that were not susceptible to symbolization.

References

Freud, S. (1948). *Our Attitude toward Death.* Madrid: Biblioteca Nueva, Tomo II.

Kristeva, J. (1995). *The New Diseases of the Soul.* Madrid: Catedra.

Lacan, J. (1971). The logical time and the assertion of anticipated certainty. A new Sofism. In: *Structuralist Reading of Freud.* Siglo veintiuno. Editors Mexico.

Milmaniene, J. (2006). *The Subject's Time.* Buenos Aires: Biblos.

Poisonnier, D. (1999). *The Drive of Death. From Freud to Lacan.* Buenos Aires: New Vision.

CHAPTER EIGHT

The first narrative, or in search of the dead father

Rosine Jozef Perelberg

> "Can one tell—that is to say, narrate—time, time itself, as such, for its own sake? That would surely be an absurd undertaking . . ."
>
> (Thomas Mann, *The Magic Mountain*)

Time and space are central dimensions in psychoanalysis, indissolubly linked to each other. The enigma of one's origins, the beginnings of desire, sexuality, and loss introduce the dimensions of space, time, and phantasy. They constitute the fundamental questions that human beings have asked about themselves since the beginnings of time, the answers to which ultimately provide a view of the individual that is not constituted solely in terms of linear development. The individual in Freud's formulations is de-centred and ruled by various temporalities, most of which escape his conscious awareness.

In observing a game played by his eighteen-month-old grandson, Freud noticed that the child threw a cotton reel and said *fort* (disappeared), and then pulled it back and said *da* (found). This is understood by Freud as the attempt to master the comings and goings of the mother. It is in the space created by the absence of the object that a sense of time is instituted and the activity of fantasy takes place. Recent discussions of this game have stressed how the child is indeed throwing the cotton reel inside the cot and thus, perhaps, also exploring the nature of his own disappearance from the mind of the mother. Who is she with, when she is gone? The beginnings of the awareness of time, linked to the comings and goings of the mother, are also connected to the awareness of the existence of the father. In the space that is thus constructed, the beginnings of the Oedipal situation are also being created. The father is already there, as a presence in the mother's mind, in her desire for him. Time, space, phantasy, and sexuality are completely intertwined. In the analytic situation, this is represented in the comings and goings of the analyst, and the beginnings and ends of sessions; weekends and holiday breaks become a metaphor for this very first narrative that is filled with our patients' desire.

Freud's paper "Mourning and melancholia" (1917b) indicates that the absence of the object opens the space for the beginnings of thinking and, one can suggest, of time. The individual identifies with the lost object and internalizes it in an ambivalent way (with love and hatred at the same time). In "The loss of reality in neurosis and psychosis" (1924e), Freud states: "But it is evident that a precondition for the setting up of reality is that the objects have been lost which once brought real satisfaction" (1924e, *S.E., 19*). In "Negation" (1925b), Freud puts forward the formulation that thinking begins "when the omnipotent control over the subjective object is shattered". The absence of the object, while inaugurating space, is also connected with the beginnings of time. The simplest narrative contains the story of an object that left and then came back. Is the dimension of loss, enquires Laplanche, "co-extensive with temporalisation itself?" (1999, p. 241). The centrality of the notion of absence and waiting in the structuring of the mind cannot be underestimated. In the analytic process, this will find its echoes in the silence and the waiting of the analytic attitude.

These ideas are central to Bion's thinking; using a mathematical metaphor, he pointed out that the geometrical concept of space derives from an experience of "the place where something was". If this concept is to be used to characterize mental phenomena, the concept of space in dreams designates the place where the lost object was, or the space where some form of emotion used to be (Bion, 1970, p. 10). Thus, it implies that an object's place has been lost. This mental space, as a thing in itself, is unknowable, although it can be represented by thoughts.

In the French psychoanalytic literature, the concept of desire is also related to an object that was known and then lost. Desire, thus, refers to an absence, and belongs to the same field as phantasy and dream (Mannoni, 1968, p. 111). That which is longed for

> only comes into existence *as an object* when it is lost to the baby or infant . . . Desire persists as an effect of a primordial absence and it therefore indicates that, in this area, there is something fundamentally impossible about satisfaction itself. [Mitchell, 1982, p. 6]

Two temporal axes permeate Freud's work: on the one hand, the genetic, that articulates development with the biological dimension of the individual's life, and, on the other hand, the structural, present in Freud's various models of the mind. In clinical practice, one is constantly alternating between these two dimensions. A moment of understanding of the *transference* might lead to a structural insight in relation to phantasies about the internal objects: mother, father, siblings, or aspects of the self. *My suggestion is that the interpretation of unconscious phantasies constitutes a link between the past and the present, the genetic and the structural.* In the analytic process one is always moving between the *there and then* and the *here and now,* two axes that are cut across by *the derivatives of the unconscious phantasies in the transference.* In this process, the present reinterprets the past in terms of *après coup,* although the past also contains the seeds that will lead to an understanding of the present, albeit with no sense of predetermination.

Green (2002a) has suggested that the therapeutic encounter takes place at many levels. The patient tells a story: of their origins, of their family or parents. The analyst is silent, paying receptive, suspended attention, facing the patient's free associations. In the current situation, the patient's past conflictual knots are reactivated. Affects, sexual life, professional life, social relationships: all these make an indissoluble whole, like a piece of music in which the analyst may pick up the themes and variations, and outline the contours of the Oedipus complex of childhood. The setting only has value as a metaphor for another concept (such as dreams, the incest taboo, parricide, maternal care, etc.). It is the derivatives of these universal questions that are enacted in the here and now of the analytic process.

In the metapsychological papers, Freud introduces the concept of primary phantasies (*Urphantasien*) that are there from the beginning, but can only be "reactivated", so to speak, in the life of each individual in terms of *après coup* (see Freud, 1917b, 1916–1917, 1918a; Laplanche & Pontalis, 1968; Perelberg, 2005, 2006; Perron, 2001; Steiner, 2003). Laplanche and Pontalis view these primary phantasies, as structuring experience. Green (2002b) suggests the notion of *the disposition to re-acquisition*. Primal fantasies are reactualized through individual experience.

What are these primal phantasies? Freud suggests that they are castration, seduction, and primal scene. Later, one could also include the Oedipus complex. In this chapter, I examine another phantasy that I see as an important organizer of psychic life, that of "a child is being beaten". I consider especially a transformation of this phantasy that Kristeva has more recently presented, which is the phantasy of "the father beaten to death" (Kristeva, 2008). Inspired by Kristeva's formulations, I suggest that the phantasy of "a father beaten to death" becomes an important achievement in the analysis of some men, as an expression of the constitution of their sexual choice and masculine identification. It is in the construction of this phantasy in the analytic encounter that the male patient finds his own temporality in the chain of generations.

The theme of the killing of the father permeates Freud's writings, from *Totem and Taboo* (1912–1913) to *Moses and Monotheism* (1939a). Freud oscillates between hypothesizing, on the one hand, that this was a real event that took place in the distant past and was repressed, but nevertheless preserved in the unconscious, and regarding the "event" as a myth, on the other hand (Godelier, 1996). Thus, a paradox is presented: the killing of the father is, in Freud's view, the requirement for the creation of the social order which, from then on, prohibits all killings. The father, however, has to be killed metaphorically only, as the exclusion of the father lies at the origin of so many psychopathologies, ranging from violence to the psychoses and perversions (see Perelberg, 2008, 2009).

These ideas are crucially linked to the centrality of the Oedipus complex in Freud's formulations, which constitutes the first, basic, symbolic structure, and includes a network of concepts such as the murder of the father, the setting up of the ego ideal, superego, desexualization, and sublimation (Green, 2004, 2008, p. 28; also Kohon, 2005). The Oedipus complex retrospectively retranslates earlier experiences in terms of *après coup*. These ideas contrast with Klein's formulation: it is the mother (or her loss) that is at the origins of symbolization; the father is an "appendage" of the mother, and the penis becomes a poor substitute for the breast (see Kohon, 1999, p. 16; and also Kristeva, 2001). For Freud, the father is crucial, as a presence in the mother's mind, but essentially as the third element that institutes the prohibition of incest in the relationship with the mother. This formulation needs also to include the child's desire. Green states,

> His essential role in the structuring of the mother–child relationship stems from the place the father occupies in the mother's mind. More precisely it depends on how she situates him with respect to the Oedipal phantasies of her own childhood. [1992, p. 134; also 2004]

This implies, first, the centrality of the triangular constellation; and second, a complex model of temporality in which the past is constantly being reinterpreted and constructed under a new light, and of different temporalities that co-exist at the same time, making up a structure.

In this chapter, I suggest that the phantasy of "a father beaten to death" and its transformations emerges for certain patients as a *result* of the work of analysis and becomes a potential appropriation of the (symbolic) father; i.e., the father as the third. This contrasts with other configurations that I have also encountered in my analytic practice, of beating daydreams or conscious fantasies that tend to constitute a foreclosure in the development of the internal relationship with the father. It also contrasts with clinical configurations where the son is actually violent towards the father (Perelberg, 1999).

I now give a detailed example of a sequence of sessions in the analysis of a young man that has given rise to some of the thoughts discussed here.

Clinical example

Mauro: background information

At the time of starting his analysis Mauro was in his early thirties, married with two children. He was the Head of the English Department at a secondary school; some time into his analysis he applied for, and got, the post of headmaster. He was in analysis for several years, and the sessions I will present took place in his fourth year.

Mauro's family comes from South Africa. The whole family came to England when Mauro was still a baby. The father left his mother when he was very young. The mother brought up her two children alone, and set up a small catering business. She seems to have been an admirable woman, devoted to her two children, and Mauro has memories of being very attached to her as a child. The children had never seen their father since he left them. The mother told them that he had been violent towards her, and the older brother remembers such scenes.

Wednesday

Mauro had had a car crash the previous week. His car was a write off but he was not harmed. His insurance company will pay for a new car. We had done some work on the significance of this accident in the analysis. On the days preceding the accident I had experienced Mauro as uninterested in things in general. This included a certain apathy connected with work and what he will be facing next term as the newly appointed headmaster. In the countertransference, I had felt a mild lack of interest too. I had addressed this in terms of the approaching break and his feelings of precariousness connected to it. In the previous session, after the car crash, I had formulated an interpretation about his longing for a protective father/analyst. Campbell's work (1995) about the role of the father in a pre-suicidal state had come to my mind at that moment, and I had thought about the links between his apathy, my momentary retreating from him, and the accident. The session I am about to narrate follows that interpretation.

> Mauro comes in and lies on the couch. He says that he found a car in Cornwall through the Internet. It is in good condition. It belongs to a policeman, and this inspires confidence.

He then says excitedly that he had a dream (he has not remembered a dream for some weeks).

"We were in a car. I was a passenger, and I was with my family, my mum and brother. We were teenagers and were driving along. Then we arrived at a street market. There were costumes from these characters in *Star Wars*, guard uniforms with masks. They were somehow lined up, but bits were missing from each of them. In one there were no boots ... We stopped in order to get out. I then realized that we were all wearing long trench coats. (When I was a child we used to go to a market in the village nearby, and I used to buy clothes there. I used to love rummaging among things, it was quite incredible the rubbish that could be bought there. Some very nice things, too. I used to get these long coats that were very warm and practical for the winter. I loved them.)

"In the dream, we were wearing these coats. Then, in the third bit of the dream (the first bit was in the car, the second was in the market), in the third bit, people wandered off, and I wanted to go to the toilet. I went and opened the door. There was an old man and he had his penis out. I went past him and he seemed to come alive and left. There was no excitement about it, a sense of disgust rather."

(Silence)

"The bit about Star Wars reminds me of someone who just started to teach at the school these past months. Chiara has recently bought a toy of this character Darth Vader for her son ..."

"Chiara came to my office yesterday. The school building is undergoing some work and scaffolding has now been set up that reaches the window of our office. As I arrived at school I saw this young builder coming to the scaffolding, and I thought that he looked really attractive ..."

(Silence)

"Darth Vader turns out to be the father of one of the characters in the film. He is the authority, but also evil. These guards are clones of this character. They all dress like him. In my mind they are quite inhuman; these characters with masks, in the dream there are the shells, these empty uniforms ..."

Analyst: "You are all dressed in these long trench coats, like when you were a child ... like Darth Vader, and there are all these bits missing; the uniforms are empty inside ..."

Patient: "I remember as a child wanting to join some sort of uniformed institution, like the police or fire brigade. [He himself makes the link.] Like the guy I am buying the car from, the police." (Silence)

(Silence)

"I was thinking of Carla's situation (another colleague at work). She really had a breakdown and had to stop working. I wonder how she is feeling now, as everything was so public, so exposed ... The headmistress did not deal well with the situation, did not pick up what was happening before when Carla asked for sick leave, she could have protected her ..."

Analyst: "Perhaps you feel that you, too, are looking for a father/ analyst who can be a protector, and not represent the dark forces, or be disgusting. Who would not let you feel ashamed, expose yourself, like the man in the toilet does; or like the headmistress who leaves Carla so exposed in front of her colleagues ..."

(Pause)

Patient: "It reminds me of my own father, who never looks for his children [or *looks at* the children, I think of the mask he had referred to, and the absence of a face to look at ...]."

Silence. Mauro then thinks of a scene that he had observed between his sister in law and her baby on the day before:

Patient: "The mother and baby were playing a game of two cars that were intertwined. The baby would hold one car and scream, shaking it with excitement. The mother, Hélène, would hold the other car, pull gently towards her. He would let it go, and reach for it again and take it back." (Silence) One could talk about it in a metaphoric way, say that this game expressed their bond, their link, the passionate desire between them ...

(Silence). [I think of his excitement at the beginning of the session in telling me the dream.]

Analyst: "Perhaps this is what it feels like when you bring me a dream."

Mauro smiles ...

It is time to stop. [I have a thought, that the world of play between mother and baby can be experienced as a refuge/regression from the world of the dark, perverse, forces, attributed to the father. Yet, at the same time, it also feels like the place from where things can be explored.]

Discussion

At the beginning of the session, the dream brings with it the expression of different temporalities: as a child/teenager with his mother and brother. What follows begins to give an indication of the search for the father, for a protective father, and yet what he encounters is a castrating/castrated/empty father. It is interesting to note that the father appears in the third part of the dream, so that the element of thirdness is potentially expressed in this way, almost in anticipation.

Mauro then tells me that he always wished to join something that involved a uniform as he grew up. Yet, there is a repetitive indication of something that is missing in each of the uniforms. Feelings of emptiness and loss are linked to his father's absence from his life, my absence in the situation of the car crash, an absence that is experienced and expressed in the impotence Mauro presented at the beginning of his analysis. It is a father who is not there to look at his children, as a witness to the relationship with the mother. The empty uniforms convey to me a sense of a negative of the father, a non-father. This seems to me a father who still does not have a space in his mind in a position of thirdness.

The reference is to the past of childhood and also the past of the analysis, in its beginning, all this evoked in the here and now of the analysis. The image of the father that emerges is that of Darth Vader, and the evocation of *Star Wars*. There is the reference to trenches and war. This evil father, who is perverse and exposes his power in the toilet, is also the longed-for father. The disgust experienced in the dream seems to also be a reaction to the longing for this father.

I also thought that the reference to the young, attractive builder is an evocation of the homosexual erotic longing that we have done quite a lot of work on throughout the analysis. The homosexual pull is a potential solution for the experience of conflicting feelings towards the father. An interpretation that addresses this longing puts Mauro in contact with the disappointment with his own father, and pulls him regressively to a scene between mother and baby that is exciting, and excludes the father.

At the Thursday session Mauro told me about the game between mother and baby that had excited him. He said:

"As a child you can just throw the toys out of the pram if you don't want to play . . . I was thinking after the session yesterday that Hélène put these two toys behind her; Roberto (the baby) was then pulling her shirt about. He could not find them and then she lost him. He went off. It is like me giving up in the past before I tried hard to do it . . ."

Analyst: "Hiding the plastic toys evokes what feels too difficult at the moment with the break. The baby loses the excitement and curiosity and goes away. Like you felt you had lost interest in the sessions, or that *I* lost interest in you."

Friday

Mauro comes and lies on the couch:

"I had a dream last night.

"There was a group of men and one of them was found to have done something: broken the law or something like that. The other turned on him, beating him, kicking him unconscious. Then a woman appeared, she was with this man who was beaten. She didn't immediately run to him. She tried to argue with the leader. She was trying to ingratiate herself with him, promising something. Her partner was still on the floor. It was unclear if she was abandoning him or trying to ingratiate herself to help him. The leader goes off and she goes with him. I walk back, and on my way back there is a Turkish patisserie with this sweet with spinach and feta cheese. There was fantastic food there. I walk past it and think that I am away for the night and I now know where to get something to eat. I was thinking about this group of men. Ordinarily you don't get kicked on the floor for breaking the law. They were part of some primitive or lawless society, where brutal justice operates. When this woman tries to ingratiate herself, there is something sexual. It really bothered me.

"When I woke up, I could not remember what the man in the dream had done wrong. I thought I had known it. I am really pleased about this dream, as I had to work hard to remember it. First I remembered the patisserie and then the rest of the dream. It was my route into the dream. Yesterday, I had to deal with two important issues. Both involved meeting the heads of department of two small schools who are interested in merging with us. One was founded the year I was born. I met with the Head of English, Marie. We were going to have coffee. I am almost there and she phones that she is late, and changed the place we were supposed to meet. After that it was like a detective story. Marie kept phoning and changing places. The purpose of the meeting was for us to meet each other and get to know each other a bit, before we get down to business. She led me on this wild goose chase. In the end she led me to a park and eventually she turns up. She is very bright, but a touch mad. I listened to her for a couple of hours in the rain. It was an unsettling meeting.

"Then I met with another head of department from another school . . ."

Pause.

"The guy I was going to buy the car from decided to sell it to somebody else. I am really disappointed. I am looking at something else. It is a fantastic car. When Christine came home last night I said I was looking at porn on the internet. They are such beautiful things, these cars . . . I was kicking myself that I had not taken time off to go down to see the car . . . He really let me down . . ."

Analyst: "It makes me think that perhaps the man in the dream was being kicked for letting you down . . ."

"I did feel kicked yesterday by these two people. I cannot afford to waste time like this. I thought how angry I am . . .

"The image that comes to mind is *The Magic Mountain*, by Thomas Mann. This is a group of civilized people! When I was writing my dream I wrote Ucs, that he was kicked unconscious. The savagery of the Ucs. What can account for my anger?"

Analyst: [I was thinking about the allusion to a perverse, seductive/destructive, timeless, atmosphere in the here and now that is like watching porn. I say:] "*A* man is being beaten . . . Perhaps the woman who watches is also me. You feel vulnerable, at the mercy of this mad woman and the sadistic man. This woman is either passively watching the scene or trying to seduce you not to be angry and violent. It reminds me of the car crash last week when you felt so vulnerable, but also so angry, ultimately the feeling that I let this happen to you."

"I did feel utterly humiliated lying there after the accident . . ."

(There is a noise outside in the street.)

"This is so annoying. Lying here, I am powerless. I can't get up and shut the window. I do feel vulnerable."

Analyst: "Like the small schools that are being taken over by a big school . . . At the same time, you also feel angry for your wish to merge with, and submit to the volatile woman who leads you on this wild goose chase (like me over the break) or to the man doing the beating . . ."

Patient: "A way to get to know people is to see what they do to you, to make yourself available. Marie made me go through these streets; I was at her mercy . . ."

(Silence for a while)

"I got home early; we had time together with Christina, which was great. We ate pesto pasta. It was really nice, like in the patisserie . . ."

Analyst: "Maybe this is the other side of your experience here. It is also like being in the patisserie."

(Silence)

"I am thinking that Christina used to have this boyfriend before who could really offer her a lavish lifestyle. She met him in Turkey and they were together for several years . . ."

(Silence). It is time.

Discussion

The Friday session starts with the narrative of a scene in which a man is being beaten. Several configurations are present at the same time. Mauro seems to experience my interpretations as either sadistically "beating him unconscious" or as seducing him into not being angry with me. In trying to seduce him, I would be ignoring the violence of the beating that contained both sadistic and masochistic identifications. Mauro identifies with different positions at the same time: he is the violent man doing the beating, as well as the man who masochistically submits to the beating. He is also the observer of the scene that becomes condensed into "a man is being beaten". The associations of the week establish a link between the man who is either beating or being beaten to the father, Darth Vader. In the passive position, the scene is homosexual, with the mother as an onlooker. This felt to me to be a transformation of the crash scene.

I had noticed the use of the French word "patisserie" that he said had been "the point of entrance" into remembering the dream. In the past he had associated French with me, because of the French books he had seen in my consulting room. It seemed to express the way in which it is the experience in the analytic process that gives access to, is the "portal", so to speak, into the unconscious. It reminded me of the end of the Wednesday session, and the description of the erotic play between mother and baby. This I understood as expressing his experience of the here and now in the transference. It is from the experience of play and pleasure, which of course also contains ambivalence, that other areas of his experience, other sources of anxiety, can be explored. These gain representation in the dream. Winnicott suggests that the progressive experience of disillusionment in the relationship with the mother leads to the activities of remembering, reliving, fantasying, and dreaming, to the "integration of past, present and future" (Winnicott, 1971, p. 12; 1972).

One can start the narrative of these sessions at any point, as there is not only a linear account, but a structure that may be comprehended. At the Wednesday session, in the first part of the dream, the teenagers were in a car with the mother. In the second part, they arrive at a market and a search is taking place. The third part of the dream involves the male figure and his penis. In the sequence of associations, and retrospectively, it is possible to understand that the search is for a father. The reference to age might hint at the life cycle, the emergence of sexuality, and the longing for an identificatory masculine object. There is a reference to a "non-father" in the empty uniforms, referring to an absence that is indeed a fact in Mauro's life. The father Mauro encounters in the toilet is exhibitionistic and perverse. He is supposed to have the phallus, "the force", but his excitement provokes disgust, which is also the counterpart of desire, of the erotic homosexual longing for him, as a precursor towards a wish for identification with him.

The father has been missing in Mauro's life, present only as an aspect of the mother's desire. His absence was also connected with his violence towards the mother, which makes it difficult for Mauro to mobilize both his aggression and his desire. If the father is the third that brings a sense of boundary to the relationship between mother and son, the experience may be that mother and son, analyst and patient, are locked in an embrace from which there is no escape. After all, Clavdia Chauchat, from *The Magic Mountain*, represents erotic temptation, lust, and love, all in a degenerate, morbid form. She is one of the major reasons for Castorp's extended stay on the magic mountain. She is the female promise of sensual pleasure as hindrance to the male's action. Chauchat's has also feline characteristics: her last name is derived from the French *chaud chat* (*hot cat*), and her first name includes the English word *claw.*

In the transference, the analyst represents both the father as an expression of the evil forces, alternating with the mother, provider of the world of play, desire, and seduction. This can be experienced as too much excitement and forbidden desire, which leads to feelings of guilt and to the beating scene of the Friday session. In this, the mother appears more clearly also as a sadistic seductress. In the succession of sessions, one can comprehend the phantasies of castration, seduction, and primal scene. They are given representation in the various scenes in the dreams that, in turn, reflect the experiences of the analytic process itself. The analyst is father, mother, child, seductress, and protector, in an alternation of identifications expressed in the vicissitudes of the transference and countertransference. Through this, the analysis is always expressing a triangular constellation, in which the father is always present in the analyst's mind.

The phantasy of the "father being beaten to death"

"A child is being beaten: a contribution to the genesis of sexual perversion" appeared in 1919, at a period of transition between Freud's models of the mind. As Catherine Chabert indicates, the text's intention was to consider the phantasy *A child is being beaten* as one of the seduction fantasies, and, in addition, to describe the paradigmatic developments involved in the production of this phantasy (2005). At the same time, it featured the "infantile" representations of masochism, anticipating works still to come before the publication of *Beyond the Pleasure Principle* (1920g), and "The economic problem of masochism" (1924c), which establishes the link between love and punishment, excitation and pain. The shift between the scenes of the beating phantasy, Chabert suggests, is a fundamental movement of the analysis, "a way of opening up positions of identification in movement" (2005, p. 226) between activity and passivity, sadism and masochism, representations and actions.

The emergence of the phantasy *A child is being beaten*, Chabert suggests, does not occur in all analyses. However, Chabert argues that this phantasy is one of the translations of the seduction phantasy linked to the primal scene, for

> it presents the full set of characteristics of originating fantasies: visual, even panoramic, support, essential for giving them a form, and the passive position assigned to the subject both in the primal scene and castration. [2005, p. 228]

It is paradigmatic of the analytic encounter, indicating the range of the identificatory processes available in it.

"A child is being beaten" is central to Freud's theoretical development and suggests a link between masochism, femininity (in both men and women, I would suggest), and the guilt feelings engendered by incestuous desires towards the father, desires that are repressed and reconstructed in the analytic process.

In Mauro's dream and associations, one can identify the transformations of the beating scene. The mother is perverse; she watches or seduces. The father does the beating or is being beaten.

Kristeva has suggested that the fantasy of "a child is being beaten" may be seen as

> representing the beginning of individuation, that decisive time when the subject constitutes himself, starting with his *sexual choice*, and then as speaking identity in the ternary structure of Oedipal kinship. I, male or female, excluded from the primal scene, look for my place between mother and father, in order to both mark out my difference, and to find my place among the ties, inseparably those of love and speaking, erotic and signifying. [2008, p. 177]

A transformation of the scene is that of a "father beaten to death", which Kristeva has proposed in a recent paper, a new reading of *Totem and Taboo* under the lens of an interpretation of "A child is being beaten". Kristeva suggests that the fantasy of "a father beaten to death" lies at the foundation of Christian faith, leading to a resexualization of the ideal father. In the light of clinical experience, *I have now come to think that it is a crucial elaboration in the analysis of many male patients*. Perhaps it is more important when the external father has not been sufficiently present to be symbolically murdered or beaten to death. The dream of beating the father will become an achievement of the analysis, as an expression of the realization of the ambivalent feelings towards the father.

In a previous paper, in connection with the analysis of another patient, I identified this phantasy, expressed in a dream and the associations in the situation of the transference (Perelberg, 2007, 2008). I related the scene of beating to the analytic situation and suggested that there is a

> masochistic, erotic dimension of the primary relationship (to the mother that) is evoked in all relationships and is an intrinsic part of the analytic setting. This idea brings trauma to the centre of the analytic experience. [2008, p. 145]

Freud's view presents a paradox in that, if the killing of the father is the requirement for the creation of a social and psychic order, the father, however, has to be killed metaphorically only, as the outcome of the Oedipus complex and the origin of the super-ego. The phantasy of the father "being beaten to death" becomes an important achievement of the analysis. It opened up the pathway to a stronger masculine identification, to the mobilization of feelings of rage and violence, as well as the potential capacity to use aggression in a more creative way. One is reminded here of the Rat Man, when he knows the pleasure of sexual intercourse for the first time: "This is glorious! One might murder one's father for this!" (Freud, 1909d, p. 201).

Darth Vader: a father is killed

The reference to *Star Wars* at the Wednesday session is also evocative of the phantasy of the murder of the father or "the father is being beaten to death". Perhaps the huge success of these films is connected with the fact that they contain within them the structure of a universal unconscious phantasy. It is also worth reminding the reader of the main plot, not only because of the main phantasy, that of the "beating the father to death", but also because of the movement between present and past and the retranslation of the past in terms of *après coup* that we, the public, were exposed to in the films.

Luke Skywalker and his friends wish to kill and destroy the evil Darth Vader, who, unknown to Luke and to all of us, is also Luke's father. The fact that he is the father will only be revealed some years later, in a second film that discloses this fact. Even then, there is some ambiguity about the veracity of this information, and it is only nearly twenty years later, when a prequel to the trilogy is made, that one learns, *après coup*, the "facts" about Luke's origins. Twenty-two years after *Star Wars* was released, a second trilogy, a prequel to the original trilogy, began to be released (Lucas, 1977–2005).

This prequel trilogy starts thirty-two years before, and follows the upbringing of Anakin Skywalker, who is believed to be the "Chosen One" foretold by Jedi prophecy to bring balance to the Force. In the remainder of the prequel trilogy, Anakin falls to the dark side. Anakin and Padmé fall in love, and eventually Padmé becomes pregnant. Anakin soon succumbs to his anger, becoming Darth Vader. Vader participates in the extermination of the Jedi Order, culminating in a light-sabre battle between himself and Obi-Wan. After defeating Darth Vader, Obi-Wan leaves him for dead—but Darth Vader is saved shortly after and put into a suit of black armour that keeps him alive. At the same time, Padmé dies while giving birth to his children, who are twins. The twins are hidden from Vader and not told of their true parentage.

The original trilogy begins nineteen years later, as Vader nears completion of the massive Death Star space station that will allow him to crush the rebellion which has formed against the evil empire. Obi-Wan begins to teach Luke about the Force, but is killed in a showdown with Vader during the rescue of Leia, Luke's twin sister. Luke seeks to train as a Jedi, but is interrupted when Vader lures him into a trap. Luke confronts Darth Vader in a light-sabre duel in the carbon freezing chamber. Luke escapes being frozen, and the duel moves to the edge of Cloud City's central wind tunnel. Luke is defeated when Vader *severs his right hand at the wrist* and sends both the hand and his light-sabre flying into the abyss. Vader takes advantage of Luke's state of weakness to tell him that he is his father. After Luke cries out in disbelief, he chooses to fall down the wind tunnel instead of accepting Vader's offer to "rule the galaxy as father and son". Later, he is saved by his sister and taken to a medical ship, where he is given a robotic hand.

Luke learns that he must face his father before he can become a Jedi, and again confronts Vader. The son defeats the father in another light-sabre duel and is able to convince him that there is still some good in him. Vader succumbs to his own injuries, and freedom is restored to the galaxy. The killing of the father frees the son.

This film evokes the myth of the Dead Father (see Taylor, 2008). It can also be seen as an illustration of Kristeva's ideas about the "father beaten to death". The father needs to be beaten so that the boy can grow up and find his place in the chain of the generations.

It is interesting to note that one of the prominent elements of *Star Wars* is the "Force", an omnipresent form of energy that can be harnessed by those with the ability to do so. It is described in the first film as "an energy field created by all living things [that] surrounds us, penetrates us, [and] binds the galaxy together". The Force allows users to perform supernatural deeds and can also amplify certain physical traits, such as speed and reflexes; these abilities can be improved through training. While the Force can be used for good, it also has a dark side that expresses hatred, aggression, and malevolence. The force becomes identification with the father's powers. In the film, there is no room for identification, a substitute, or sublimation; the force has to be stolen, and the father needs to be killed, as the Oedipal myth dictates.

Conclusion

Freud's myth, in which the father needs to be murdered in order to be erected as the third, seems to me to be a mythical account of the process of growing up, where the parents need to be destroyed by the adolescent in order for him to grow.

The analytic process, in so many ways, recapitulates the narrative of the dead father complex. First, is the analytic setting itself not establishing the place where the law of the father is expressed? In a previous paper I stated,

> I feel, however, that when the analyst formulates interpretations—of whatever kind—she is *inaugurating* something for the patient, independently of the content of the interpretation. The analyst introduces differentiations and separations into a territory previously more chaotic and undifferentiated. The theories present in the analyst's formulations are thus not there, present in the mind of the patient, available to be uncovered, but become constructions made by both the analyst and the patient in the analytical process. In this

> process, the analyst is by definition creating the paternal function and breaking up the phantasy of a fusion with the mother. When the patients reject these interpretations, in the process of working through, are they then not, by definition, attempting to reinstate the fusion with his mother? [Perelberg, 1999, p. 105]

The elaboration of the Oedipus complex and the relinquishing of one's incestuous phantasies place the individual in a temporal dimension. The object needs to be gained in order to be lost and then re-presented, as Freud indicates with the analysis of the fort-da game. This requires facing the desire to kill the father, or "beat the father to death", as has been possible in Mauro's analysis.

I have now encountered this phantasy of a "father is being beaten" in several analyses of male patients (see also Perelberg, 2007) and suggest that it is a crucial achievement in the analysis of many male patients. It initiates the process of mourning that allows them to initiate their experience of their own temporality and historicization.

Acknowledgement

I am grateful to Routledge and The New Library of Psychoanalysis for permission to quote from Chapters One and Two from my book *Time, Space and Phantasy*, Routledge, London, 2008.

References

Bion, W. R. (1970). *Attention and Interpretation*. London: Tavistock.

Campbell, D. (1995). The role of the father in a pre-suicide state. *International Journal of Psycho-Analysis, 76*(2): 315–323. Also in: R. J. Perelberg (Ed.), *Psychoanalytic Understanding of Violence and Suicide* (pp. 63–73). London: Routledge, 1999.

Chabert, C. (2005). Clinical and metapsychological thoughts derived from "A child is being beaten". In: R. J. Perelberg (Ed.), *Freud: A Modern Reader* (pp. 224–233). London: Wiley.

Freud, S. (1909d). *Notes upon a Case of Obsessional Neurosis. S.E., 10*: 153–249. London: Hogarth.

Freud, S. (1912–1913). *Totem and Taboo. S.E., 13*: 1–161. London: Hogarth.

Freud, S. (1916–1917). *Introductory Lectures on Psychoanalysis. S.E., 15–16.* London: Hogarth.

Freud, S. (1917b [1915]). Mourning and melancholia. *S.E., 14*: 237–258. London: Hogarth.

Freud, S. (1918a). *From the History of an Infantile Neurosis. S.E., 17*: 3–123. London: Hogarth.

Freud, S. (1919e). A child is being beaten: a contribution to the study of the origin of sexual perversions. *S.E., 17*: 175–204. London: Hogarth.

Freud, S. (1920g). *Beyond the Pleasure Principle. S.E., 18*: 7–64.

Freud, S. (1924c). The economic problem of masochism. *S.E., 19*: 159–170. London: Hogarth.

Freud, S. (1924e). Loss of reality in neurosis and psychosis. *S.E., 19*: 183–190. London: Hogarth.

Freud, S. (1925b). Negation. *S.E., 19*: 235–239. London: Hogarth.

Freud, S. (1939a). *Moses and Monotheism. S.E., 23*: 1–139. London: Hogarth.

Godelier, M. (1996). Meurtre du père ou sacrifice de la sexualité. In: M. Godelier & J. Hassoun (Eds.), *Meurtre du Père Sacrifice de la sexualité: approches anthropologiques et psychanalytiques* (pp. 19–52). Paris: Arcanes.

Green, A. (1992). *La Déliaison*. Paris: Belles Lettres.

Green, A. (2002a). *Idées directrices pour une psychanalyse contemporaine*. Paris: PUF.

Green, A. (2002b). *Time in Psychoanalysis: Some Contradictory Aspects*. London: Free Associations.

Green, A. (2004). Thirdness and psychoanalytic concepts. *Psychoanalytic Quarterly, LXXIII*: 99–135.

Green, A. (2008). The construction of the lost father. In: L. Kalinich & S. Taylor (Ed.), *The Dead Father: A Psychoanalytic Enquiry* (pp. 23–46). London: Routledge.

Kohon, G. (1999). *No Lost Certainties to be Recovered*. London: Karnac.

Kohon, G. (2005). The oedipus complex. In: S. Budd & R. Rusbridger (Eds.), *Introducing Psychoanalysis: Essential Themes and Topics* (pp. 166–180). London: Routledge.

Kristeva, J. (2001). *Melanie Klein*. New York: Columbia University Press.

Kristeva, J. (2008). A father is beaten to death. In: L. Kalinich & S. Taylor (Eds.), *The Dead Father: A Psychoanalytic Enquiry* (pp. 175–187). London: Routledge.

Laplanche, J. (1999). *Essays on Otherness*. London: Routledge.

Laplanche, J., & Pontalis, J.-B. (1968). Fantasy and the origins of sexuality. *The International Journal of Psychoanalysis, 49*: 1–18.

Lucas, G. (1977–2005). *Star Wars: Motion Picture,* Lucasfilm Ltd and 20th Century Fox, California [written and directed by George Lucas].

Mannoni, M. (1968). *Freud: The Theory of the Unconscious*. London: Verso.

Mitchell, J. (1982). Introduction (1). In: J. Mitchell & J. Rose (Eds.), *Feminine Sexuality—Jacques Lacan and the École Freudienne* (pp. 1–26). London: McMillan.

Perelberg, R. J. (1999). A core phantasy in violence. In: R. J. Perelberg (Ed.), *Psychoanalytic Understanding of Violence and Suicide* (pp. 87–108). London: Routledge.

Perelberg, R. J. (2005). Unconscious phantasy and *après coup*: from the history of an infantile neurosis. In: R. J. Perelberg (Ed.), *Freud: A Modern Reader* (pp. 206–223). London: Whurr.

Perelberg, R. J. (2006). Controversial discussions and *après-coup*. *International Journal of Psychoanalysis, 87*: 1199–1220. Also in Perelberg, R. J. (2008). *Time, Space and Phantasy*. London: Routledge.

Perelberg, R. J. (2007). Space and time in psychoanalytic listening. *International Journal of Psychoanalysis, 88*: 1473–1490. Also in Perelberg, R. J. (2008). *Time, Space and Phantasy*. London: Routledge.

Perelberg, R. J. (2008). *Time, Space and Phantasy*. London: Routledge.

Perelberg, R. J. (2009). Murdered father; dead father: revisiting the Oedipus complex. *International Journal of Psychoanalysis* (forthcoming).

Perron, R. (2001). The unconscious and primal phantasies. *International Journal of Psychoanalysis, 82*(3): 583–595.

Steiner, R. (Ed.) (2003). *Unconscious Phantasy*. London: Karnac.

Taylor, S. (2008). Prologue. In: L. Kalinich & S. Taylor (Eds.), *The Dead Father: A Psychoanalytic Enquiry* (pp. 9–18). London: Routledge.

Winnicott, D. W. (1971). *Playing and Reality*. London: Tavistock.

Winnicott, D. W. (1972). The use of an object. *International Journal of Psychoanalysis, 50*: 711–716.

CHAPTER NINE

The destruction of time in pathological narcissism*

Otto Kernberg

Introduction

As Elliot Jaques (1982) pointed out in his overview of psychoanalytic views of the experience of time, it is important to keep in mind the difference between objective time as a scientific concept characterized by the uniformity of linear intervals as defined by the units of measurement of time, on the one hand, and the subjective sense of time, that has very different characteristics, on the other. The subjective experience of the duration of time is irregular and depends on multiple psychological factors.

Throughout the life cycle, a remarkable yet gradual change occurs in the subjective experience of the duration of time. The multitude of early experiences that bombard the infant and small child gradually settle into longer cycles between the past and the future, such as, for example, the long time in between weekends, and the endless time between birthdays, thus taking on a quality of "endless time", the correlate to the naturally assumed permanence

*Previously published in 2008 in *International Journal of Psychoanalysis, 89*(2): 299–312.

of childhood. With developing growth and maturity, and a more predictable succession of tasks and personal investments, cycles of past experience seem to accelerate. The expectation of future developments, that are now more firmly embedded in consciousness by the individual's own life trajectory, planning, and task investments, matched with active work towards the transformation of such a projected future into the present, decreases the subjective experience of the duration of time so that it seems to be passing more rapidly. There is a clearer sense of what to expect in the future, and a sharp linkage between past experience and its expected repetition. The sense of acceleration of the passing of time increases with age, and becomes a significant conscious experience in old age (Hartocollis, 1983). Now time "flies".

Happy moments, "stellar experiences", while seeming to pass too quickly, none the less build up as happy memories, creating a sense of life lived intensely, that extends the sense of duration of time across the life span. The opposite development characterizes traumatic experiences. Severe trauma has multiple influences on the subjective sense of time, depending on the nature and duration of the traumatic experience. In the case of acute, brief situations when the trauma is the product of wilful aggression, there will be an almost intolerable sense of extension of time during the traumatic experience itself, with a fixation to the trauma that, by repetitive "flashbacks", extends the subjectively experienced duration of the trauma. The long-range effect of this situation leads to a "time stood still" quality related to reverberating unconscious processes that reduce, retrospectively, the experience of time, particularly that of time lived after the traumatic experience. Thus, for example, a couple who were assaulted, robbed, and controlled with threats to their life over a period of hours had a grossly distorted subjective experience of extension of the duration of the event, with a post traumatic stress disorder, fixation to the trauma over a period of many months, and a retrospective sense of shrinkage of the time after the trauma over one to three years. It was "as if it happened yesterday".

For extended periods of wilfully induced traumatic circumstances, for example, racial persecution, concentration camp imprisonment, or extended periods of physical or sexual abuse, the effect is even more powerful: the dominance of the unconscious consequences of the traumatic situation reduce the capacity for significant new investments and, with it, the loss of a generation of new experiences that otherwise would enrich the experience of passage of time.

In these last examples of cognitive and traumatic influences on the sense of time, the function of memory of an accumulated life experience becomes important. This is, in fact, a complementary dimension of the experience of time, the sense of time lived intensively. The more significant the investment in meaningful and gratifying relationships and activities, the more the moment seems to fly by, but, by the same token, there grows a sense of time having been lived and an enrichment of the total life experience. If, to the contrary, such a meaningful commitment to investments in work, art, social engagements, and, as we shall see in more detail, to intimate relations, is missed, experience of life lived shrinks, and life itself may seem to be near its end, accompanied by a frightening sense of the brevity of time lived.

The pathologically persistent dominance of primitive dissociation or splitting operations (which characterize the early stage of development) in the syndrome of identity diffusion characteristic of severe personality disorders leads not only to the threatening reactivation of dreaded bad experiences that have to be avoided or denied, but also to a search for idealized ones that, in turn, cannot be reactivated fully because the reality of the experience of object relations does not ever totally fulfil that idealized world. All this reduces the possibility of integrating new experiences, and condemns the individual to repeatedly relive a subjectively unchanging world of dreaded and fleetingly idealized experiences. As a result, under such pathological circumstances, repetition compulsion condenses the sense of lived time; new experiences cannot be integrated normally; traumatic situations are recreated that require constant attention to the immediate environment and do not permit new, gratifying experiences to build up a significant past. Repetition compulsion has many sources and functions, but one consequence relevant here is the implied denial of the passage of time: "nothing has changed, the repetition indicates that time is frozen". The duration of time shrinks, in contrast to what happens with the deepening of emotional relationships that characterizes normal identity (the depressive position). This shrinkage of time is even more accentuated in the case of patients with narcissistic personality disorders. Here, the devaluation of significant others as a defence against unconscious envy is reflected in the dismantling of internalized object relations. The pathological grandiose self is experienced in isolation, and self-esteem regulation is dependent on external admiration from others, rather than on the security of an internalized world of significant object relations. The failure to develop significant object relations results in a chronically empty internal world, depleted of emotionally deep and meaningful experiences, that condenses, retrospectively, the experience of time: nothing memorable has happened in the past, except the ongoing efforts to shore up self-esteem and confirm the grandiosity of the self. The narcissistic patients will often find themselves "waking up" at age forty, fifty, or sixty with a desperate sense of years lost.

In contrast, integrated whole object relations permit the build-up of a lived past, the sense of duration of time lived expands, and a desired, imagined future extends it further. Internalized relations with loved and gratifying others determine the time-framed memories of interactions, real and fantasized, in contrast to the rigid and static memories of stereotyped others with whom no joint history was built up, and no duration of time invested in such interactional sequence is established. Guilt and reparation of past aggression, mourning the lost idealization of the past, the reinforcement of the sense of a good self by gratifying preconscious and unconscious relationships with significant others fill up life and time. Life, then, is experienced as intense and hopeful; the future holds the expectation for ongoing good experiences, all of which reinforces self-esteem, zestful optimism, and the affirmation of life.

Identity itself develops further in this context, as identity of childhood expands throughout adulthood with the internalization of identifications with significant parental objects at different stages of their life, so that a future can be projected in which one's identity is partially modelled upon the identity of an older generation of strongly invested parental figures and mentors. The acceptance of one's own past and the resolution of early Oedipal and pre-Oedipal conflicts permit identification with one's own children, so that the total life experience is enriched by a projected future and a reliving of an accepted past in its creative re-creation and modification with a younger generation. Identity, in short, simultaneously expands towards future and past, and that, in turn, enriches life experience in the sense of life lived intensively, while subjective time expands accordingly.

These developments are relevant for the psychology and pathology of the ageing process. The expansion of identity implies the capacity for identification with past and future generations, their interests, struggles, and experiences, and provides a sense of continuity of life. In contrast, the failure of this process to form normal identity with its corresponding time dimension, together with a sense of the shrinking of time in the ageing process referred to before, may bring about an increased fear of death. Narcissistic personalities frequently experience, in later decades of life, a sense of not having lived sufficiently, that life has gone by without leaving traces of the past. The experiences of shrinkage of time, in these cases, may bring about an intense and growing fear of death, a sense of unfairness of the brevity of their life as they experience it. This fear is also related to infantile fears of abandonment and loneliness, and a deep feeling of the senselessness of life, which predominate when there is an absence of investment in love, work, ideals, children, and values. The functions of ideology, religion, art, and culture as vehicles for creation of values, as well as of human communication and a sense of the continuity of humanity, cannot be internalized fully under circumstances of identity diffusion and the structural dominance of a pathological grandiose self. In contrast, investment in one's own lived history and in the history of those one is involved with, and the transcendence of this investment into a general sense of historical continuity, provide a reinforcing context to the sense of living and of having lived a full life.

A particularly painful experience of "lost time" may become part of the mourning process, both normal and pathological. Guilt feelings stirred up in the mourning process over not having fully lived the time that was available with the loved person who has been lost (a normal expression of the depressive position) is experienced with much more severity in pathological mourning. In narcissistic personalities, this may take the form of a complete absence of normal mourning, a denial of guilt feelings that cannot be tolerated because of their potentially frightening intensity, or else the emergence of paranoid behaviour reflecting the projection of intolerable guilt feelings.

Normal mourning, as Melanie Klein (1940) observed, always involves guilt feelings as an essential aspect of the activation and reworking of the depressive position. The death of a beloved person always illuminates, retrospectively, the infinite number of lost occasions and possibilities of expanding the intimate communication with the lost person, the time that could have been lived together and was not, the feelings of love that were not expressed; in short, the total value of that relationship that could not be actualized under the impact of daily routines not overshadowed by the awareness of death, of a final separation.

Rabbi Moshe Berger (personal communication) has stated that all loving relationships are finite, while their loss through death initiates an infinite absence, which only now illuminates all possible aspects and values of the relationship, in contrast to the necessarily limited awareness of them during the deceased loved person's life with us. Only the infinite absence permits us to become fully aware of all the implications, meanings, and possibilities of a finite relationship. Such awareness heightens the regret and guilt over the "waste" of objective time with the loved one and, under optimal circumstances, will induce, in the mourning person, a heightened subjective experience of the time lived with the lost love object. This process is often blocked in the case of narcissistic personalities, for whom the awareness of guilt, regret, and dependency threatens to overwhelm the pathological grandiose self.

Hartocollis (2003) has described the regressive effects of timelessness induced by free association, and the counteracting effect of the precision, the consistency of duration, and the regularity of the analytic sessions. The unconscious denial of the dependency on the analyst characteristic of narcissistic patients transforms the relationship with him into a static, self-indulging focus on internal processes, fantasies, and wishes that are not linked to the time-generating mutuality of a changing object relation. This might not be perceived by the analyst over a period of time. Thus, the regressive effect of the method of free association is significantly increased in the treatment of narcissistic personalities: here, the timelessness of analysis lends itself to express one specific aspect of narcissistic pathology. The invitation to free associate, with its explicit discouragement of "prepared agendas" and related moves towards action, is often misinterpreted as an invitation to passivity that narcissistic patients unconsciously translate as a projection of all responsibility on the analyst, and a defiant expectation of gratification from him . . . and his defeat.

The dominant pathology of the time experience of narcissistic personalities derives from the destruction of their internalized world of object relations, a result of the development of a pathological grandiose self that incorporates real and idealized representations of self and others. This leads to the devaluation of others who otherwise would generate envy, and the resulting lack of internalization of gratifying relationships with significant others brings about an impoverishment of the internal world, with its absence of time-bound meaningful interactions in depth. In replacement of such experiences, these patients experience the need for immediate gratification from external sources, be it admiration from others, triumph and success as recognized in the external world that confirms their narcissistic superiority, or, if none of these is available, an escape in depersonalized sexual relations, drugs, alcohol, or other sources of immediate excitement (Kernberg, 1984, 1992).

This stark picture is modified by the fact that many narcissistic personalities do not suffer from such a total destruction of the world of internalized object relations, and are able, for example, to obtain narcissistic gratification while making an important investment in relation to their children, on to whom they project their own narcissistic needs. A particular task, talent, or social function at which they excel may generate social recognition and gratification, but also provide them with an intrinsic pleasure derived from an investment in such an object. There are, however, cases where, even in the face of apparently well-preserved social functioning, the absence of investment in relations with others, and the resulting sense of internal emptiness, are significant features. These cases are often well compensated as long as success at work, a profession, or other social endeavours provide them with adequate narcissistic gratification, but suffer greatly with illness, retirement, loss of power or recognition. The sense of a lack of lived experiences, with the implicit shrinkage of the duration of time, is an important aspect of their sense of emptiness and fear, an intuition of the waste of time.

In psychoanalytic treatment, narcissistic patients typically evince their defences against dependency on the analyst by an unconscious devaluation of what they receive from him/her, thus warding off envy of the therapist they need. This process may bring about a stubborn resistance against change, as interpretations fall on sterile ground. Here, there is an active, unconscious effort that "nothing should happen", the direct expression of a self-destructive triumph over the analyst that reinforces the regressive effects of free association in them. It is as if time was standing still in the analysis, and patients typically complain at such points that nothing is helping them. This shrinkage of time, however, may correspond to a still deeper transference development, the unconscious desire that time does indeed stand still. One aspect of the function of the grandiose self is precisely the denial of the passage of time, the fantasy of eternal youth, and the very denial of death as an ultimate threat to their grandiosity. While the fantasy of eternal youth and the denial of death may be a universal manifestation of normal infantile narcissism, in the narcissistic personality it becomes grossly exaggerated, an aspect of the pathological grandiose self that interferes seriously with a realistic adaptation to the objective passage of time. All this reinforces the function of the unconscious destruction of time in the analytic relationship of such patients, the assertion of their invulnerability to the influence of the treatment, the defeat of the analyst's work as an expression of unconscious envy of him. The emptying out of the narcissistic patient's life experiences during analysis, in fantasy, becomes a triumph over the analyst's capacity to influence them. Green (2007) has pointed to the function of repetition compulsion, when it is employed as a form of "murder of time", as an expression of the death drive. This certainly applies to some cases of narcissistic personalities.

One narcissistic patient, a successful businessman, entered psychoanalytic treatment, four sessions a week, because of the incapacity to commit himself to a satisfactory relationship with women. After a lengthy period of indecisiveness regarding marriage, he did marry a woman following a brief infatuation, which proved as dissatisfying and "boring" as all other relationships. He presented himself as a shrewd and superior businessman, but was easily upset by minor slights or lack of consideration on the part of others, contemptuous of friends and family, particularly of the family of his wife. He was consistently surprised by the intense social life that his wife had with her family, while he felt only resentment and devaluation towards various members of his own family, including his parents. It struck him that he had great difficulty in remembering not only names, but also faces and the very existence of people whom he had met throughout his high school and college years, and later on during the time of travel abroad and vacations. He had a perfect memory for all aspects of his business, was an expert in a certain historical style of furniture, and could judge authenticity in origins and styles. He was superficially friendly—as long as he felt admired—and, although numerous business associates had attempted to establish a more personal relationship with him, he was unable to involve himself in anything beyond direct business negotiations. With a strong sense that his marital life was not satisfactory, he had attempted to establish relations with other women, but soon discovered in them limitations, became bored with them, and went on to the next. His only major symptom, other than a chronic degree of anxiety, was a fear of death and related hypochondriacal concerns, and anxious ruminations over his getting old without ever having lived. He had a clear sense that time had passed him by, that he had not really lived, that his death would mean having been cheated out of life before he had a chance to live it meaningfully.

His success as a major athlete in the past was related to his interest in sports but, once his own active participation was no longer possible, he devalued the interest of the sports he had been involved with as well. He had described his mother as a dominant, overly anxious person, hypochondriacal, concerned over his, her only son's, health, but he had no memory of any interactions with her other than her controlling his appearance, eating patterns, behaviour and health. Father, a rather unsuccessful businessman, was despised by my patient, and he saw the purpose of his own life in not letting himself be exploited and defeated as he felt his father had been in his business interventions.

In this patient's analysis, the development of a typical narcissistic resistance against involvement in the transference was expressed as a "matter of fact" dominance of immediate, realistic concerns in this patient's life, with almost total absence of any phantasy material, and no reference to, or curiosity about, the analyst. He saw analysis as an opportunity to resolve his business-related preoccupations, and tended to dismiss the interpretations of the analyst as bookish and theoretical. At the same time, he always felt pressed for time, everything had to be resolved rapidly, and he resented the "timelessness" of his sessions, namely, the analyst's patiently listening to free associations instead of indicating courses of action the patient might pursue. He distrusted the value of the analyst's comments as much as the importance of what might come to the patient's mind. Nothing was going on in his analysis, he proclaimed triumphantly, while rejecting most interventions of the analyst. And yet he seemed willing to come to sessions punctually without question. He reiterated his conviction that he had no feelings for the analyst: analysis was a special business dealing . . .

Thus, fear over the emptiness of time not spent in business considerations or practical life situations coincided with the emptying out of meaningful interactions in his sessions: he was always in a hurry, and nothing seemed to happen in his emotional life. Gradually, focus on his hypochondriacal fear, and his fear of death began to uncover his dread over a lack of anything emotionally moving in his life, and eventually intense envy of the analyst as somebody he feared had a rich life experience.

In one session, after complaining at length that nothing was changing in his life, that he was still bored with his wife and dissatisfied with the lack of excitement in everyday experiences, he suddenly laughed and said that he had an image of me sitting puzzled in my chair, unable to help him, yet condemned to be sitting like that for an infinite number of hours. He went on to say that he, actually, was young enough to find new exciting experiences, while I was ageing, and time was passing me by while I was stuck in a questionable profession. I pointed out to him that, in that fantasy, he remained eternally young with unlimited possibilities, while I, in addition to failing him, would be struck with old age, and with good reasons to envy his youth.

The patient became anxious, wondered whether I was angry with him, and, later in the session, realized that, in fact, he had felt angry in the past session because of a new book authored by me that he had discovered on my desk. I then pointed out that it became clearer why, at this point, the thought of lack of progress and waste of time in his analysis had not upset him . . . Only after many months of this development, after working through his intense envy of the analyst, emerged wishes for an idealized relationship with a powerful father, forbidden because it contained wishes for a homosexual relation that he was terribly afraid of. Eventually, the tolerance of his homosexual feelings brought back wishes for friendship in his early adolescence, and fears of being rejected by another boy whom he was deeply invested in. Now more lively memories of his past could be elicited in his associations. But, behind longing for a previously repressed, dependent relationship with a good father emerged his deep disappointment in mother, and distrust and hatred of women. He gradually became aware of a profound resentment of all women because he thought they were so self-sufficient and did not need anybody else, that it would be dangerous to look for other than a temporary sexual relation with them. The conflict around unconscious envy of the analyst now emerged as an expression of his hatred of mother, the deep distrust of depending on her, and the resentment of her power to soothe and to mistreat him. Now the unconscious triumph over me, by asserting my impotence to touch and to change him, could be explored in the hours. Only towards the end of the treatment did Oedipal issues of competitiveness with me become prominent, a sense of triumph over me because he was significantly younger than me, and, in this context, an awareness of the fear of death as an expression of the projection of his unconscious rivalry wishes to eliminate me, and a sense of Oedipal defeat.

In the context of these developments, this patient began to experience a profound regret for missed opportunities, friends whom he had rejected, women whom he had not been able to appreciate, and, above all, his neglect and devaluation of his wife's capacity for investment and love that had made him feel terribly envious and inferior. He now began to enjoy his daily life with her that he had taken for granted before, as well as relations with friends and family. He found new interests in travel with his wife, and lost the chronic sense of emptiness together with the fear of death as a confirmation of the uselessness of his life. He was no longer chronically rushing from one encounter to another, and could enjoy, for the first time, a contemplative attitude toward his wife and friends.

Another narcissistic patient, also in psychoanalytic treatment, four sessions a week, a man in his mid-twenties with an extraordinary talent as a painter and as a specialist in the Spanish language, had spent years in both these fields, earning early recognition and applause, and stimulation to continue in one of these careers, but was unable to engage in the work required in order to progress technically or to depend on other experts in order to develop his own technique at a more mature level. His envious resentment of those who would have been able to teach him made him devalue both areas of his expertise and, in the end, abandon them with resentment of those who were successful in them, and the painful awareness that he had not achieved anything nor obtained any gratification in areas of those talents in which he had invested more than ten years of his life. He consulted because he felt uncertain what profession to select at that point, and, in fact, was working in a subordinate function in a field totally unrelated to his learning experiences. He deprecated his present work, but could not decide what else to do. He was depressed and neglectful of his appearance. Rather soon, in the treatment, he recognized a sense of superiority, of expectations to rise to the top without the effort of a long road ahead of him. Here, a sense of destruction of time emerged at a point when he realized that the fantasy that he was going to be eternally young and promising, and that, therefore, nothing was lost by avoiding the learning opportunities that would have required, as he saw it, a humiliating sense of not being perfect. The fantasy that he was young and had "all the roads to the future" open to him broke down when confronted with the reality of the success of all those whom he had considered inferior to him, and who now were making creative changes in their lives. For a long time, his image of the analyst was one of a passive, "resigned" mediocrity, who could only do the same thing all the time. The awareness that the analyst's interpretations reflected active, creative efforts to understand and to help the patient came as a serious blow; the fact that time advanced, was finite, and could be wasted and lost, was a painful new experience.

Still another narcissistic patient, in psychoanalytic psychotherapy, three sessions a week, a woman from an aristocratic European environment who only wished to get involved with leading members of that social group, devalued and dismissed all those lovers who were not part of that group. Now in her late forties, she began to experience the wish to get married, and for the first time began to question the haughty way in which she had treated men, and her triumphant enjoyment of her seductive capacity without having been able to relate in depth to any of the men she was involved with. Her sense of an empty, wasted life triggered a severe depression that brought her to treatment. That sense reflected her successive abandonment of work and interests she had not been able to sustain because of the envy of those who were ahead of her, and the endless repetition of her disappointments in all the men she met—mostly narcissistic personalities whose perceived grandiosity had attracted her at first—and the lack of meaningful, ongoing relations in depth. She expressed very concretely her terror and sense of loss that she had become forty years of age without having had a sense of really living that long: where had the time gone between a turbulent adolescence through twenty years of routine parties and social engagements?

The most severe cases in which destruction of time becomes dominant are those who almost wilfully destroy their opportunities, and manage, eventually, to attach themselves to highly destructive partners, with whom they establish a sadomasochistic relationship that, in turn, tends to further reduce their possibilities and potential. Couples of this type may hold on to an eternal repetition of self-defeating fights and mutual accusations, thus neglecting the impoverishment of their life through this fixation to a destructive object. The absence of the sense of the passage of time may be expressed in the unrelenting fixation to a relationship in which the patient binds another person to himself or herself, in an unconscious need to maintain a fantasy relationship that, while destructive to both parties, replaces a real one, sometimes over many years without any real content or interaction. In some cases, what looks on the surface as being in love with an unavailable person turns out to be a disguised self-condemnation to loneliness and emptiness as time seems to collapse in the permanent uncertainty of their lives. Ruminating over months—and years!—over whether or not they should have engaged in a certain love relationship may dramatically obscure their awareness of the passage of time.

In the analysis of patients where the destruction of time is an expression of narcissistic denial of the reality of the passage of time and severely restricts the possibilities of life, unconsciously the patient may repeat the pattern of destructiveness of object relations in the transference by maintaining himself in an analytic situation that, on the surface, is supposed to treat his difficulties, but that, unconsciously, is used to maintain the equilibrium of narcissistic emptiness and triumph over a parental figure, the analyst who is trying to help the patient get out of this bind. The unconscious use of the destruction of time as a triumph over the analyst while also expressing the fantasy of an available eternity of life to the narcissistic patient may, initially, escape the analyst's attention; the patient may, unconsciously, tease the analyst with apparent changes that prove their lack of substance throughout time. In the early stages of the analysis of such patients, what grabs one's attention is the superficiality of relations with significant others. The patient may describe the personality of people he is involved with in rather behavioural, even categorical, fashion, but it is almost impossible for the analyst to get a real image of such other persons. This feature is, of course, quite typical for all narcissistic personalities, who have enormous difficulty in an assessment in depth and in the development of significant relationships with others, but here the degree of trivialization of the descriptions, and the endless repetition of the same content reaches a very high degree, so that it is as if the patient was relating to robots with repetitive behaviours that, for some strange reason, fascinate the patient.

Efforts to raise questions about this kind of information are typically met, not only with the patient's sense of puzzlement, fear of being criticized, and the need to defend the "realistic" way in which he relates to others, but with opening up the transference analysis of similar developments with the analyst, who may be perceived as being interested in the patient for the analyst's own benefit or his wishes to be a successful therapist, but without any real interest in the patient. The lack of reflection of the patient outside the sessions on what is being discussed in the hours, is striking. Any active effort of the analyst to provide some degree of depth to the work acquires the characteristic of a "first session", as if the analysis is just starting at that point. This situation also reflects the patient's subjective timelessness in the hours, as if objective time spent in the hours is magically going to help him even if, in fact, nothing inside the patient really changes. These are also patients who, precisely because nothing is happening in the hours, easily get bored or even fall asleep, and, of course, use whatever information they have about "active therapies" of one kind or another to demand a change in the analyst's approach. This feature of the transference may have a discouraging effect on the analyst: it is as if the analysis is starting all over again and again.

The destruction of time may take many forms. Some patients seem to "learn" everything they hear from the analyst, associate to the interpretations in ways that may appear to be confirmatory of them, including the emergence of new, relevant material, conveying an emotional reception of what evolves in the session. But nothing evolves after the session. They maintain perfect memory of what was said, and of their reaction, but do not evince any further curiosity about it, so that, weeks later, the same material may be presented as if it were the first time in which it came up.

At times, the patient questions what had evolved during a particular session, but without sharing these questions with the analyst for quite some time. Other people with whom the patient shares what transpired in the session will disagree with the analyst's observations. Or simple "forgetting" occurs, particularly of central points focused upon in the sessions. There are patients who experience a depersonalization during the sessions, as if they were listening and reacting to issues involving somebody else, even being able to communicate this experience to the analyst without any change in it.

The lack of these patients' reflections on their thoughts and feelings, on the analyst's comments, and on their own incapacity to reflect on what they were helped to become aware of in the sessions are a consistent aspect of their relation to the analyst and his interpretations. They may become aware of intense envy of the analyst, and, while their envy becomes conscious, their efforts to neutralize it by a lack of response to the analyst's efforts to help them remain unconscious.

The analyst's countertransference may be the dominant instrument signalling an alarm reaction faced with the stagnation of the treatment. The patient's incapacity to depend on the analyst may gradually threaten to undermine his commitment to the patient. Aggression in the countertransference may be the only indication of massive projective identification of resentful rage of an envied parental object on the part of the patient. The subliminal expression of such countertransference reaction in interpretive comments may be triumphantly interpreted by the patient as the analyst's "loss of patience", and therefore, the analyst's problem!

Lengthy stalemates may develop, with the analyst's oscillating internally between efforts to find new ways to deal with the stalemate in interpretive fashion, and the impulse to set limits to the time in which such a total lack of movement of the treatment should be tolerated. The question, whether the secondary gain of the defeat of the analyst's efforts reinforces the patient's commitment to the denial of the passage of time sufficiently to condemn the treatment to failure, may become an acute concern for the anayst. An internal rejection of the patient, reflecting a projective counter-identification with the patient's defensive denial of his needs for dependency may complicate the countertransference.

The solution to this complication is the analysis of the very unavailability to the patient of the analyst as a person, as somebody who thinks, reacts, reflects, and is touched by what is going on in the sessions, and can be generalized to the same unavailability to the patient of everybody else. Very often, what can be found at a deeper level is the unconscious identification of the patient with a parental object that treated the patient as an object without an internal life, so that the patient treats the analyst as he was treated by his parental object, and expects the analyst, of course, to treat him in the same way. If the analyst treats him very differently from the way he was treated in the past, a new world opens up that the patient may experience as painfully illustrating the contrary nature of the terrible deprivation from his past. The pain over a frustrating, empty, past childhood is very difficult for the patient to tolerate, and may stimulate envy of the analyst for not having been subjected to such a terrible past experience. To begin to depend on the analyst under such conditions may be experienced by the patient as a terribly humiliating defeat, and this reaction may induce regressive cycles of withdrawals and disrupted periods of dependency on the analyst.

Because psychoanalysis is such a long-term treatment, and these patients may be strangely satisfied by regularly coming to the sessions in spite of no discernible progress; all the while giving the appearance of freely associating to varying types of situations, feelings, and understandings, they may unconsciously gratify the analyst's wishes for a dependent relationship on the part of a patient who, apparently, makes no demands and has no expectations for change other than those sudden outbursts of interests in "active treatments". Not infrequently, after a period of time, the analyst may be tempted to move the treatment into a more supportive direction, thus, in turn, gratifying the patient's need to receive narcissistic gratification by this personal "trainer". After a period of time, the development of uneasiness and guilt feelings in the analyst over the growing awareness of an absence of progress may, in fact, paralyse him regarding any efforts to "start all over again", that is, to risk the activation of a new "first session", in which the unavailability of the patient to the analyst and of the analyst to the patient can be explored.

Naturally, the general "safety measures" that serve to indicate whether or not the analysis is progressing may alert the analyst to what is happening, and help him face this extremely difficult situation. These measures include the question of what the patient does with the interpretations, how they affect the patient from one session to the next, what changes, if any, are occurring in the transference and in the countertransference. In this regard, the awareness and the analysis of chronic countertransference developments become relevant in these cases, where the analyst gradually discovers that nothing seems to be happening except the objective passage of time, without the patient being disturbed by this changeless passage of time.

It is important, faced with a patient where active destruction of time seems to be a significant expression of unconscious destructiveness and self-destructiveness, that the analyst tolerates an experience in himself of impatience in every session, while mustering a great degree of patience to deal with the situation over an extended period of time. This impatience within each session might be reflected in a consistent, active effort to deal with the patient's lack of deepening the present object relationship, and exploration of what it means in terms of transference and countertransference developments. Such impatience runs counter to a misunderstood overextension of the principle of analysing "without memory or desire", as Bion (1967) had formulated it in the past. That healthy principle of a neutral analytic attitude runs the risk of being co-opted here into the destructive developments of the transference. I believe that the psychoanalyst has a responsibility to attempt to help the patient, and this responsibility includes an optimal use of the time of each session, and is not to be considered a form of *furor sanandis*.

The destruction of time as an expression of the destruction of the internalized relations with significant others may find a dangerous collusion in an analytic approach in which the healthy aspects of the analyst's patience and the long-lasting nature of psychoanalytic treatment are contaminated by the unconscious identification of the analyst with a certain culture of psychoanalytic institutes to extend all educational and supervisory processes to the greatest length possible. The analytic culture in some institutes tolerates and fosters analyses of a duration extending beyond ten or even twenty years, and discourages candidates from "impatience" to graduate and become fully independent. Such cultures may reinforce the delay in diagnosing the narcissistic pathology in analytic treatment we are considering. This may lead to a quiet acceptance of an apparently endless analytic process in which there may be intense fireworks, but no real deepening of the emotional investments related by the patient and actualized in the transference–countertransference bind. An unconscious collusion between a narcissistic patient and the analyst regarding the "timelessness" of analysis may be particularly dangerous under these circumstances. The destruction of time may, at times, be another expression of the syndrome of the "dead mother" described by Green (1993), the unconscious identification with, and link to, a severely depressed mother experienced as a dead one, that presents itself usually, however, in patients who look much more severely ill on the surface. Typically, in the patients discussed by Green, the devaluing behaviour, the manifest derogatory indifference, and the stubborn rejection of everything the analyst has to offer are quite evident from the beginning of treatment. In the case of narcissistic patients, to the contrary, a surface friendliness, a social easiness, an apparently much more successful social life, and even an apparently intimate one may dominate, so that the destruction of time that goes with a deep unavailability of significant object relations may take time to come to the concerned attention of the analyst.

In conclusion, the destruction of time in narcissistic pathology may serve various functions: the expression of unconscious envy of the analyst, the denial of the unavailability of the grandiose self to any change, the simple consequence of the unconscious destruction of internalized object relations that would fill objective time with meaning, and whose absence shrinks subjective time and condemns the patient to the experience of an empty life.

References

Bion, W. R. (1967). Notes on memory and desire. *Psychoanalytic Forum, 2*: 272–273, 279–280.

Green, A. (1993). *On Private Madness*. Madison, CT: International Universities Press.

Green, A. (2007). From the ignorance of time to the murder of time: from the murder of time to the misrecognition of temporality in psychoanalysis. *Psychoanalysis in Europe, 61*: 12–25.

Hartocollis, P. (1983). Time and the life cycle. *Time and Timelessness, 15*: 215–226.

Hartocollis, P. (2003). Time and the psychoanalytic situation. *Psychoanalytic Quarterly, 72*: 939–957.

Jaques, E. (1982). *The Form of Time*. New York: Crane, Russak.

Kernberg, O. F. (1984). *Severe Personality Disorders: Psychotherapeutic Strategies*. New Haven, CT: Yale University Press.

Kernberg, O. F. (1992). *Aggression in Personality Disorders and Perversions*. New Haven, CT: Yale University Press.

Klein, M. (1940). Mourning and its relation to manic-depressive states. *International Journal of Psychoanalysis, 21*: 125–153.

CHAPTER TEN

Hindu concepts of time

Satish Reddy

> "I am time grown old to destroy the world,
> Embarked on the course of world annihilation:
> Except for yourself none of these will survive,
> Of these warriors arrayed in opposite armies.
> Therefore raise yourself now and reap rich fame,
> Rule the plentiful realm by defeating your foes!
> I myself have doomed them ages ago:
> Be thou the mere instrument, Left-handed Archer!"
>
> (The *Bhagavad Gita*)

I will start with the simple and intuitive notion that the past determines the present and the present determines the future. This is as true in Hinduism as it is true in psychoanalysis. The Sanskrit word for time is *kala*. Time in Hinduism is considered to be cyclical, rather than linear. Time is viewed on a macroscopic cosmological level, consisting of vast cycles that repeat eternally. On the individual level of human existence, however, time functions both linearly and cyclically. Four critical concepts of Hinduism—*Samsara, Karma, Dharma* and *Moksha*—introduce us to the concept of time in Hinduism. These concepts broadly determine Hinduism's existential stance and are the starting point, framework, and context within which a Hindu views, structures, and lives his life. They are both the fundamental presuppositions of Hinduism and the emotional driving force of Hindu religiosity.

The story of Indra and the parade of the ants

This story from the *Brahmavaivarta Purana*, wonderfully and eloquently retold by Zimmer (1974), introduces us to the conception of time in Hinduism as well as the critical concepts of *Samsara, Dharma, Karma,* and *Moksha,* which together outline the Hindu existential position and draw the trajectory of an individual life through the vast cycles of time.

Indra, king of the gods, slew the demon holding the waters of creation in his belly. By killing the demon, Indra allowed the natural course of creation to progress and re-established the power of the gods in heaven. To commemorate his victory and the start of a new era in heaven, he employed the divine architect Vishvakarman to build a city worthy of the gods in heaven. Vishvakarman applied himself and built a beautiful city unparalled in previous heavens; yet, Indra was not satisfied. He wanted more splendid palaces and more grandeur. Indra's demands increased to the point that Vishvakarman, exhausted and frustrated by Indra's demands, sought help from the supreme deity, Vishnu. Having heard the architect's plight, Vishnu nodded, indicating that he would take care of the situation.

The next day, Vishnu, in the form of a young Brahmin boy, came to visit Indra. As it is customary to honour Brahmins—even for the gods—Indra welcomed the boy and brought him into his palace. The boy admired his palace and the new city Indra was building, and sang praises to Indra and his grand accomplishments in heaven. He then asked Indra how much longer it would take to finish the task. He told him that, indeed, no other Indra before him had built such a magnificent city.

> Full of the wine of triumph, the king of the gods was entertained by the mere boy's pretension to knowledge of Indras earlier than

> himself. With a fatherly smile, he put the question: "Tell me, Child! Are they then so very many, the Indras and Vishvakarmans whom you have seen-or at least, whom you have heard of?" The wonderful guest calmly nodded. "Yes, indeed, many have I seen . . . O King of Gods, I have known the dreadful dissolution of the universe. I have seen all perish, again and again, at the end of every cycle. At that terrible time, every single atom dissolves into the primal, pure waters of eternity, whence originally all arose. Everything then goes back into the fathomless, wild infinity of the ocean, which is covered with utter darkness and is empty of every sign of animate being. Ah, who will count the universes that have passed away, or the creations that have risen afresh, again and again, from the formless abyss of the vast waters? Who will number the passing ages of the world, as they follow each other endlessly? And who will search through the wide infinities of space to count the universes side by side, each containing its Brahma, its Vishnu and its Shiva? Who will count the Indras in them all—those Indras side by side, who reign at once in all the innumerable worlds; those others who passed away before them; or even the Indras who succeed each other in any given line, ascending to godly kingship, one by one, and one by one, passing away.
>
> The life and kingship of an Indra endure seventy-one eons, and when twenty-eight Indras have expired, one Day and Night of Brahma has elapsed. But the existence of one Brahma, measured in such Brahma Days and Nights, is only one hundred and eight years. Brahma follows Brahma; one sinks and the next arises; the endless series cannot be told. There is no end to the number of those Brahmas—to say nothing of Indras." [Zimmer, 1974, pp. 5–6]

After saying this, the boy broke into laughter. He saw a parade of ants marching through the palace. Indra, frightened and bewildered asked the boy why he was laughing. The boy said that he laughed because of the ants.

> "I saw the ants, O Indra, filing in long parade. Each was once an Indra. Like you, each by virtue of pious deeds once ascended to the rank of the gods. But now, through many rebirths, each has become again an ant. This army is an army of former Indras." [*ibid.*, p. 7]

Disguised as a boy, Vishnu's comments served to modulate Indra's hubris and his individual ego. By placing Indra into the context of the many Indras that came before him and that will come after him in the vast recurring cycles of time, Indra was reminded of the Hindu perspective that one is born, dies, and is reborn: this is the doctrine of *Samsara*, or reincarnation. The boy also told Indra that his individuality and ego were not immaterial, for what one does now and how one acts determine what the person will do and become in the future, in this life and the next. This is the doctrine of *Karma:* the Hindu law of causation and retribution. Acting morally, according to one's *Dharma*, will determine whether one becomes an ant or a god.

Hindu concepts of time

Hindu concepts of time revolve around the periodic and infinite repetition of the creation (*srsti*) and dissolution (*pralaya*) of the universe. This aspect of repetition distinguishes the Hindu cosmogony from that of the monotheistic/Semitic religions (Judaism, Christianity, and Islam), wherein the creation and the destruction of the world is strictly linear (i.e., from Genesis to the Last Judgement) (see Peters, 2003). Central to the cycle of the creation and destruction of the universe are the notions of *Kalpas* and *Yugas*. A *Kalpa* is 10,000 divine years, or ten million human years!

> According to the mythologies of Hinduism, each world cycle is subdivided into four *Yugas* or world ages. These are comparable to the four ages of the Greco-Roman tradition, and like the latter decline in moral excellence as the round proceeds. The classical ages took their names from the metals—gold, silver, brass and iron—the Hindu from the four throws of the Hindu dice game—*Krita, Treta, Dvapara,* and *Kali.* In both cases the appellations suggest the relative virtues of the periods, as they succeed each other in a slow, irreversible procession. [Zimmer, 1974, p. 13]

A trinity of gods retains the responsibility for the periodic creation and destruction of the universe. Brahma is the creator; Vishnu, the preserver; and Shiva, the destroyer. In the Hindu cosmogony, each *Kalpa* represents one day and night in the hundred-year life span of the god Brahma. Thus, one day and night of Brahma corresponds to 4,320,000 human years! These years are divided into the four *Yugas*, each successive *Yuga* declining in "moral excellence", as reflected by the behaviour of human beings who are less spiritual, more materialistic, and more self-interested by the last *Yuga, Kaliyuga*. According to Indian mathematical calculations, our current age, *Kaliyuga*, began on 18 February 3102 BC. It is important to keep in mind that there have been an infinite number of such *Kaliyugas* in the past, and that there will be an infinite number of *Kaliyugas* to come. As Axel Michaels explains:

> Every *Kalpa* lasts a thousand great ages of the world (*mahayuga*), corresponding to 12,000 god years or 4,320,000 human years, which are divided into the four ages of the world (*yuga*), and each of them has a tenth long dawn and dusk, in which Brahma (or Vishnu) rests. The whole thing is repeated a thousand fold, a hundred Brahma years or 311 billion and forty million human years. When the epoch (*para*) is over, the world declines. The coarse material will again become subtle primeval material, in which the constituents are in balance, until they are shaken—either by themselves or by a divine impulse—and the cycle of the emergence and passing away of the worlds (*samsara*) continues. [Michaels, 2004, p. 300]

Such a conception of time is indeed extraordinary, not only because of its sheer magnitude, but also because of the precision of the mathematics involved. In this conception of time, years are related to humans, gods, and the "super" divinities (i.e., Brahma, Vishnu, and Shiva), indicating that not only humans, but also gods and super gods are subject to the laws of *karma* and *samsara*; they, too, are endlessly created and pass away:

> Behind this is a deeply rooted cyclical awareness of time, which holds that life consists of an eternal return, of an eternally new expansion and contraction of the world. Thus, the *Yuga* doctrine is also known as the doctrine of the world cycle (*samsara*). Everything passes away according to these ideas; only change itself is lasting, but only the condition beyond this change brings salvation. [*ibid.*, p. 303]

For the individual, this view of time places a particular individual life in perspective. Rather than rendering any one life meaningless, it implies the importance of doing one's duty and acting morally in this life: if our actions are not "good", the *karma* from these actions will make us transmigrate eternally, and it is this scary and horrifying scenario—known as *Samsara*—and the need to escape it that motivates Hindu religious thinking and life (Biardeau, 1989).

Samsara

Samsara refers to the Hindu concept of reincarnation or transmigration: in Sanskrit, it means "to wander". *Samsara* is accepted in all Hindu philosophical systems of thought, with the singular exception of the *Carvaka* school of Indian materialism that denies the existence of a soul (Chattopadhyaya, 1990). (The *Carvaka* School, as such, also denies the concepts of *karma* and *dharma*. Their view of time is linear, and, as such, is uniquely more western.) It is fair to say that the goal of Hinduism is to escape *samsara*, and to ultimately exit the endless cycle of birth and rebirth.

In order to understand *samsara*, a brief discussion of the Hindu view of the Soul or Self is necessary. The Hindu notion of "Self" differs fundamentally and radically from the psychoanalytic concept of the "self" or "ego". This point cannot be emphasized strongly enough, as it is precisely the identification of the "ego" with the "Self" that is the bedrock problem for Hindu philosophy and psychology. According to Hinduism, it is the confusion (*avidya*) of "Self" with "ego" that leads to human suffering.

Hinduism postulates a fundamental distinction between "matter" and "soul" (here, "Soul", "Self", and "Spirit" are considered identical and used interchangeably). In *Samkhya* (which literally means "to enumerate"), the oldest Hindu philosophical system, a critical distinction is made between the Soul (*purusha*) and matter (*prakti*). The Soul is conceptualized as contentless consciousness, which becomes entangled with matter throughout its life. (For an extensive discussion of *Samkhya* philosophy, see Larson & Bhattacharya, 1987.) The reason for this entanglement is not explained. It is simply given and stated; indeed, the association is considered to exceed the grasp of human comprehension.

Interestingly, and perhaps surprising to some schooled in western traditions, *Samkhya* postulates that the mind, including intellect, consciousness, emotions, and psychological states, is ultimately a derivative and function of matter (*prakti*), in concordance with contemporary neuroscientific theories of mind. Thus, what we would view in psychoanalysis as conscious and unconscious psychic structures are, according to *Samkhya*, evolutes of matter. In explicit distinction, *purusha* (Soul/Self) is independent of matter (*prakti*): it is unconditioned, timeless, has always existed, and will never cease to exist. To say that *purusha* is timeless has subtle connotations. It is strictly true only when *purusha* is not entangled with matter. When it is involved with matter, it functions in time and attains karmic residue, but is not altered by it. This is similar to Freud's assertion that the unconscious is timeless; that is, not changed or altered by time. This is not to compare *purusha* to the Freudian unconscious, but rather to point out an interesting dichotomous parallel.

> From all eternity, Spirit has found itself drawn into this illusory relation with psycho-mental life (that is, with "matter"). This is owing to ignorance (*avidya*), and as long as *avidya* persists, existence is present (by virtue of *karma*), and with it suffering. Let us dwell on this point a little. Illusion or ignorance consists with confusing the motionless and eternal *purusha* with the flux of psycho-mental life. To say "I suffer", "I want", "I hate", "I know" and to think that this "I" refers to Spirit, is to live in illusion and prolong it; for all our acts and intentions, by the simple fact that they are dependent upon *prakti*, upon "matter", are conditioned and governed by *karma*. [Eliade, 1969, p. 28]

In Hinduism, there are complex traditions that view the soul theistically and atheistically. The *Samkhya* postulation of the soul is strictly atheistic: it is purely descriptive, and involves neither god nor divinity.

> This true and absolute knowledge—which must not be confused with intellectual activity, which is psychological in essence—is not obtained by experience but by a revelation. Nothing divine plays a part here, for *Samkhya* denies the existence of god. [*ibid.*, p. 29]

Here, revelation is not to be understood as conveyed by the divine, as revealed, as in the Judeo-Christian tradition, but as achieved through self-knowledge and self-exploration (Radhakrishnan,

1939). In Hinduism, one road to self-revelation, or self-knowledge, is via the instruction of a pupil by a guru. The parallel to the analytic situation, where self-knowledge occurs through such an interaction between an analyst and an analysand, is striking and worth noting (Reddy, 2005). Sudhir Kakar, in his work, *The Analyst and the Mystic* (1991), in the chapter "The guru as healer", discusses this point in detail.

For the Hindu, this life is one of many lives we live. What then transmigrates when we die? It is the Soul-*purusha*: contentless consciousness. After the death of a person, the Soul moves to another body. Its new destination (an ant or a god) is determined by the law of *karma*. *Karma* dictates that the nature of actions committed by a particular combination of *purusha* and *prakti* determine where the *purusha*, with its karmic residue, will go after death to once more become entangled with matter (*prakti*). This cycle repeats indefinitely until the Soul no longer attains karmic residue, and can exit the cycle of *samsara* to become pure contentless consciousness again. As the god Krishna explains to Arjuna, his pupil and devotee, in the sacred Hindu text, the *Bhagavad Gita*:

> "Never was there a time when I did not exist, or you, or these kings, nor shall any of us cease to exist hereafter. Just as creatures with bodies pass through childhood, youth and old age in their bodies, so there is a passage to another body, and a wise man is not confused about it . . . There is no becoming of what did not exist, there is no unbecoming of what does exist: those who see the principles see the boundary between the two . . . As a man discards his worn-out clothes and puts on different ones that are new, so the one in the body discards aged bodies and joins with other ones that are new." [van Buitenen, 1981, Chapter Two, Verses 11–22]

The central concept of *samsara* is that actions determine one's fate; the specifics of a person's reincarnation after death are determined by the actions of that person in this life. It is a fair and profound question to ask, then, to what extent does fate (or biology or circumstance or pure chance) determine one's actions? Essentially, Hindu teachings hold that human acts arise from a combination of given circumstances and free will. Furthermore, an individual acting under these circumstances is responsible for his actions, and these actions produce karmic residues. This notion remains controversial and represents the Hindu contextualization of the perennial conflict between free will and determinism.

The state reached when one is not reborn, after exiting *samsara*, is termed *moksha*: in Buddhism, *nirvana*. What, then, does it mean to exit *samsara*? Is to exit time? Is it to withdraw from life? No. It is to be in this life and *to act without attaining karmic residue*. Is this not a contradiction? For if we act, we attain *karma*. How then is it possible to act without attaining *karma*? The Hindu solution to this dilemma is *Karmayoga*: "disinterested action", as enumerated in the *Bhagavad Gita* by Krishna to Arjuna:

> "A person does not avoid incurring *karman* just by not performing acts, nor does he achieve success by giving up acts. For no one lives even for a moment without doing some act, for the three forces of nature cause everyone to act. He who, while curbing the faculties of action, yet in his mind indulges his memories of sense objects is called a self-deceiving hypocrite. But he who curbs his senses with his mind, Arjuna, and then *disinterestedly* undertakes the discipline of action with his action faculties, stands out . . . All the world is in bondage to the *karman* of action, except for action for the purposes of sacrifice: therefore, engage in action for that purpose, disinterestedly, Kaunteya." [*ibid.*, Chapter Three, verses 5–9]

Karma

Karma literally means action. The law of *karma* is that every action has a reaction or consequence. It is a law of cause and effect. The action may be intentional, conscious or unconscious. Regardless, it has a consequence. Every action is modulated and determined by previous actions and future actions are similarly modulated and determined by present actions. The concept of *karma* links an individual's past, present, and future and presents the continuum or context in which a person, more specifically the "ego" (*ahamkara*) exists and functions. When an action is performed, it is performed by the "ego" and the karmic residue of the action stays with the "ego". This is psychic determinism in Hinduism. We are psychophysically conditioned by actions we perform and the *karma* accumulated from these actions determines individual behaviour, both in this life and the next. The concept of *karma* is very similar and analogous to the psychoanalytic notion of psychic determinism:

> The sense of this principle [of psychic determinism] is that in the mind as in physical nature about us nothing happens by chance or in a random way. Each psychic event is determined by the ones that preceded it. Events in our mental lives that may seem to be random and unrelated to what went on before are only apparently so. In fact, mental phenomena are no more capable of such a lack of causal connection with what preceded them than are physical ones. Discontinuity in this sense does not exist in mental life. [Waelder, 1963, p. 17]

What is the "Ego" in Hindu psychology that performs actions and attains karmic residue? It is not the same as the concept of "ego" in psychoanalysis. Zimmer explains:

> *Ahamkara,* the ego function, causes us to believe that we feel like acting, that we are suffering, etc.; whereas actually our real being, the *Purusha,* is devoid of such modifications. *Ahamkara* is the center and prime motivating force of "delusion". *Ahamkara* is the misconception, conceit, supposition, or belief that refers all objects and acts of consciousness to an "I" *(aham). Ahamkara*—the making *(kara)* of the utterance "I" *(aham)*—accomplishes all psychic processes, producing the misleading notion "I am hearing; I am seeing; I am rich and mighty; I am enjoying; I am about to suffer." It is thus the primal cause of the critical "wrong conception" that dogs all phenomenal experience; the idea, namely, that the life-monad *(purusha)* is implicated in, nay is identical with, the processes of living matter *(prakti).* One is continually appropriating to oneself, as a result of the *Ahamkara,* everything that comes to pass in the realms of the physique and psyche, superimposing perpetually the false notion (and apparent experience) of a subject (an "I") of all the deeds and sorrows. [1989, p. 319]

This notion of the Hindu ego, *ahamkara,* is very important to understand, because it is the temporal construction of a person, one that defines an individual's personality, their tastes, wishes, predispositions, habits, and actions. *Ahamkara* exists and functions in present time, and it and the physical body are what die at death. The Hindu "ego" *(ahamkara)* is a function of matter or *prakti,* not of the *purusha* or "Self". The Self is unconditioned and timeless, as previously explained. It is the association of the Self with matter that gives rise to the "ego". It is precisely the identification of the Self as "ego" (*ahamkara*) that is the fundamental existential and psychological error (ignorance or *avidya*) that leads actions to attain *karma* and make the Self transmigrate indefinitely.

The Hindu doctrine of *karma* is both a psychology and philosophy of action. It is a theory of motivation, a conceptual schema to explain why we do what we do. Importantly, it is also a moral and ethical concept. The moral and ethical dimension of *karma* is an important differentiating factor between psychic determinism in Hinduism *vs*. psychoanalysis. On a descriptive level, *karma* describes and explains human behaviour. On the moral level, each action has a moral valence: actions are "good" or "bad" based upon the individual's *dharma* (their particular moral duty, not absolute, Kantian style, moral duty, but individual duty). The moral valence of one's actions determines an individual's life trajectory, temporally in this life and transtemporally in future lives. Even if we disallow future lives (i.e., reincarnation), we are still left with a powerful concept, that actions are predetermined, that they have a moral value and that the moral valence is based on *dharma* or one's specific moral duty in society. I take this point up in more detail in the next section on *dharma*. As Kakar explains:

> *Karma* influences the Hindu world image in two fundamental ways: in the Hindu's experience of time, and in the formation of his cosmology. The way in which a culture estimates and elaborates ideas of time and destiny provides insight into the psychological organization of its individual members. [1981, p. 45]

Is *karma* fate? Does *karma* annul free will? If all acts are predetermined, where does free will come into play? The question may be asked of human behaviour viewed through the psychoanalytic lens of psychic determinism. Does psychic determinism mean that we act without free will? If our actions are produced by psychic structures and forces that inexorably drive us to act in a particular way, where does that leave an individual's freedom to act? Intuitively and from common sense, we act as if we have free will. *Karma* is in part fate. But a basic tenet of *karma* is also the presence of free will; indeed, this is where *dharma* and the moral valence of actions come into play. We may be "driven" to act based on our past *karma*, but we have a choice to do otherwise. It is precisely the exercising of our free will, in the face of competing karmic residues, that underlies the moral basis of *karma*.

For the Hindu, *karma* has profound psychological import. If something bad occurs, a Hindu is apt to blame it on his previous *karma*. Resorting to *karma* to explain one's actions does not absolve the act from its moral value. How one navigates between a given set of circumstances and the options available for action are determined by *karma* and determine *karma*. This is a profoundly complex and subtle theory of action where karmic forces are multi-directional and actions are "over determined". Free will exists and operates with responsibility and consequences for the actions done.

Dharma

Derived from the Sanskrit *dhr*, which means "to hold", the concept of *dharma* is complex. It can be, and is, variously and simultaneously translated as duty, law, correct moral action, and acting in accordance with one's nature. *Dharma* refers both to individual moral duty and social responsibility (Kane, 1974). Acting according to one's *dharma* refers to actions contextualized by one's position in society (*caste*) and stage of life (*asrama*). Thus, the *dharma* or correct action for a Brahmin is different from the *dharma* of a Kshatriya (kings and warriors). In the *Bhagavad Gita*, Krishna says to Arjuna: "It is better to carry out your own law (swadharma) poorly, than another's (paradharma) well; it is better to die in your own law than to prosper in another's" (van Buitenen, Chapter Three, verse 35).

Kakar explains how *swadharma* is simultaneously specifiable and subjective:

> . . . how does the individual acquire knowledge of his *svadharma*, and thus of "right actions"? This is a complicated matter, and, as it happens, a relative one. Hindu philosophy and ethics teach us that "right action" for an individual depends on *desa*, the culture in which he is born; on *kala*, the period of historical time in which he

> lives; on *srama*, the efforts required of him at different stages of life; and on *gunas*, the innate psychobiological traits which are the heritage of an individual's previous lives. "Right" and "wrong" are relative; they emerge as clear distinctions only out of the total configuration of the four "co-ordinates" of action. [1981, p. 37]

Psychoanalytically, we may see *dharma* as analogous to the "ego ideal". *Dharma* is the force, or principle, that maintains social order by regulating and modulating interactions between individuals within and between different castes. When each individual performs his duty according to his nature, state, and stage in his life, order and equilibrium is maintained in the individual and in society. *Dharma* is traditionally depicted as a wheel (*charkha*): the wheel can be seen to rotate properly when its spokes (individuals) are properly aligned and functioning.

While *dharma* refers to duty based on caste and stage of life, it also, and more subtly, refers to acting according to one's nature. In *Yoga* psychology (for a concise and excellent introduction to *Yoga* psychology, see Dasgupta, 1930), the term *vasana* refers to the compelling deep urges in us, gathered from past fields of action that now determine our present emotional profile. *Vasana* derives from the Sanskrit root *vas*, which means "to dwell in, to abide". Note the striking resemblance of *vasana* to the psychoanalytic notion of an "instinct". The difference between a *vasana* and an "instinct" may be that *vasanas* are conditioned (and, to some extent, determined) by our prior actions (i.e., *karma*), unlike instincts, which are not conditioned by experience or time. *Vasanas* determine the specific nature of an individual. Eliade explains:

> The *vasanas* condition the specific character of each individual; and this conditioning is in accordance both with his heredity and with his karmic situation. Indeed, everything that defines the intransmissible specificity of the individual, as well as the structure of the human instincts, is produced by the *vasanas*, by the subconscious. The subconscious is transmitted either "impersonally", from generation to generation (through language, mores, civilization-ethnic and historical transmission), or directly through karmic transmigration. [1969, p. 42]

When *dharma* is referred to as acting according to one's nature, it may be more accurately described as acting despite one's nature, or overcoming one's instinctual impulses. For example, my *vasanas* (instincts) may propel me towards a certain act. But if this act is not in accordance with my *dharma*, then I must not do it. To not do something that I am inclined to do requires self-control and restraint. This produces "good *karma*", which in turn modulates and modifies my *vasanas*. This circuit repeats indefinitely and multidirectionally until *moksha* is achieved. *Dharma* may be seen as adjudication between the demands of one's mental–psychic structure and the demands of society. From a psychoanalytic viewpoint, this is analogous to the ego's action based on the demands of the pleasure principle and the constraints of the reality principle, as manifest in societal constraints, and psychologically enforced by the superego.

When we act according to our *dharma*, our actions are "good" in so far as they attain less karmic residue and take us closer to the possibility of exiting *samsara* and attaining *moksha*. Drawing a circle around social order, moral order, and divine order, *dharma* links the concepts of *karma*, *samsara*, and *moksha*.

Moksha

Moksha means release—spiritual release—and freedom and escape from the endless circuit of life–death–rebirth cycle, *samsara*. The attainment of *moksha* is the *paramartha*, or highest goal of Hinduism. *Moksha* is considered the fourth aim of life, the other three being *artha* (material wealth), *kama* (love and sensual/sexual pleasure), and *dharma* (religious, moral, family, and social duties). Hinduism considers each of these aims important and necessary for an individual. And each aim should be pursued and enjoyed according to the timetable prescribed by the Hindu scriptures. The time frame of pursuing these aims is provided by the Hindu life cycle, known as *asramas*. Kakar explains:

> Like modern theories of personality, the Hindu model of *asramadharma* conceptualizes human development in a succession of stages. It holds that development proceeds not at a steady pace with a smooth continuum, but in discontinuous steps, with marked changes as the individual moves into a new phase of life: proper

> developmental progress requires the meeting and surmounting of the critical task of each phase in the proper sequence and at the proper time. Essentially, *asramadharma* is the Hindu counterpart to man's development in relation to his society and, as I have shown in detail elsewhere, it is very similar to Erickson's well known theory of psychosocial stages of growth. Contrasting with Erickson's model, which is clinical and developmental, the Hindu view proposes "ideal" images in the Platonic sense. In outlining the stages of life and the specific tasks of each stage, the Hindu model does not chart the implications for mental health if the tasks remain unfulfilled, but emphasizes the importance of scrupulous progression from task to task and from stage to stage in the ultimate realization of *moksha*. [1981, pp. 42–43]

The Hindu *asrama*, or stages of life, prescribe when each aim should be pursued. It is not appropriate, for example, for a Hindu man in the householder stage of life (*garhasthya*) to give up his duties to his family and move to the forest to meditate on the nature of god. (It is interesting to consider if the Buddha, from the Hindu point of view, was guilty of abdicating his family and royal duties (*garhasthya* stage) to pursue the withdrawal (*vanprastha*) and renunciation (*sanyasa*) stages before the prescribed timetable.) Table 1, from Kakar's book, compares the Hindu stages of life with Erickson's psychosocial scheme.

Moksha derives from the Sanskrit root "muc", which means "to release", "set free", "let go". Philosophically and psychologically, *moksha* is seen as the *purusha* ("Soul", "Self", "Spirit") disentangling itself from *prakti* (primordial matter), and returning to a state of contentless consciousness.

What exits, and what remains, in *moshka*? It is the Soul (*purusha*). When the Soul exits time, it no longer exists in time. It can thus be said to be *timeless*. A subtlety must be noted here. *Timelessness* can imply something not being conditioned or changed by time: for instance, Freud's notion of the timelessness of the unconscious. *Timelessness* can also refer to an emotional experience, sometimes referred to as *nunc stans*, or "the abiding instant", in which, as described by Loewald:

> There is no division of past, present, and future, no remembering, no wish, no anticipation, merely the complete absorption in being,

Table 1. A comparison of the Hindu stages of life with Erickson's psychosocial scheme.

Erikson's scheme		Hindu scheme	
Stage	Specific task and "virtue"	Stage	Specific task and "virtue"
1. Infancy	Basic trust *vs.* mistrust: hope	Individual's pre-history not explicitly considered	Preparation of the capacity to comprehend *dharma*
2. Early childhood	Autonomy *vs.* shame, doubt: willpower		
3. Play age	Initiative *vs.* guilt: purpose		
4. School age	Industry *vs.* inferiority: competence	1. Apprenticeship (*brahmacharya*)	Knowledge of *dharma:* Competence and fidelity
5. Adolescence	Identity *vs.* identity Diffusion: fidelity		
6. Young adulthood	Intimacy *vs.* isolation: love	2. Householder (*garhasthya*)	Practice of *dharma:* Love and care
7. Adulthood	Generativity *vs.* stagnation: care	3. Withdrawal (*vanaprastha*)	Teaching of *dharma:* Extended care
8. Old age	Integrity vs. despair: wisdom	4. Renunciation (*sanyasa*)	Realization of *dharma:* wisdom

> or in that which is . . . the experience of eternity does not include everlastingness. Time as something which, in its modes of past, present, and future, articulates experience and conveys such concepts as succession, simultaneity, and duration is suspended in such a state. In as much as this experience, however, can be remembered, it tends to be described retrospectively in temporal terms which seem to approximate or be similar to such a state. [1972, p. 405]

The notion of *moksha* has been understood in psychoanalysis broadly within the concept of fusion states. A sizeable psychoanalytic literature has been devoted to understanding and explaining these states, going back to Rolland's (1984) correspondence with Freud regarding the Hindu saint Ramakrishna and the states of *samadhi* or timeless bliss he accessed. (For an extensive discussion of the "oceanic feeling", see Parsons, 1999.) Freud considered this an *affect*, one he named the "oceanic feeling", and conceptualized it as a fusion state: a narcissistic regression to a symbiotic connection to the primordial mother. It is important to point out the difficulty—indeed, the impossibility—of understanding *moksha* through psychoanalysis, in part because *moksha* cannot properly be understood as an *affect*, which is, within Hindu thought, necessarily material; that is, issuing from the body, and thus essentially not only physically, but *temporally* finite. To exit *samsara*, and to attain *moksha*, is to literally leave the cycle and circuit of time. The experience of timelessness of the "oceanic feeling" is fundamentally different from the notion of actually exiting time. *Moksha* is an ontological condition, not an emotional or psychological state. *Moksha* cannot be said to be experienced at all. *Moksha*, simply put, is a state of not existing in time.

Another reason *moksha* cannot be understood within psychoanalysis is because, as a theory, psychoanalysis rejects the Cartesian mind–body dualism required to admit the presence of a "Soul" or "Self" in as much as the Hindu understands it; that is, as separate from matter. In psychoanalysis, the mind is essentially and *only* material. *There is no "Soul" or "Self" as in Hinduism*. Hence, psychoanalytic attempts to understand the notion of *moksha* tend to be reductionistic, simplistic, and ultimately of little use and relevance. However, there is significant potential for cross-fertilization of ideas between psychoanalysis and Hindu thought, specifically, the Hindu systems of *Samkhya* and *Yoga*, which developed theories of the mind and unconscious mental functioning long before psychoanalysis. One interesting and fascinating area for future exploration is the *Samkhya* notion that mind, intellect, emotions, and even consciousness are all evolutes and derivatives of matter.

Conclusion

The notion of time in Hinduism, while fascinating, is overwhelming and frightening. The Hindu view of the eternal recurrence of the cosmos and the vastness of the time cycles involved makes one question the value of any individual existence (Eliade, 1954). In discussing the Hindu concepts of *samsara, karma, dharma,* and *moksha,* I hope the reader has come to understand, that, for the Hindu, it is precisely the suffering inherent to human existence and the need to escape it that emotionally motivates and drives Hindu philosophy and religion. While the Hindu conception of time can be frightening, the notion of cyclical time and eternal recurrence of the Self is also strangely reassuring and edifying. Man's place within the ages is assigned and understood, and his behaviour at each stage in life can be guided according to his *dharma*. And, by recapitulating the cycle of ages in microscopic form within each man's own tiny and finite life, a connection to the incomprehensibly immense macro-cosmos is maintained, ensuring that all individual lives retain relevance and meaning. Through eternal recurrence, the Self has the possibility of discarding its karmic residue and attaining freedom: no ordinary freedom, but the freedom to exist as unconditioned contentless consciousness; pure being which exists both in time and outside of it. In the *Maitri Upanishad* (Radhakrishnan, 1953), it is said that there are two forms of *Brahman*-time and the timeless.

> Time cooks all things,
> Indeed, in the great self.
> He who knows in what time is cooked
> He is the knower of the Veda.

References

Biardeau, M. (1989). *Hinduism: The Anthropology of a Civilization*. R. Nice (Trans.). Delhi: Oxford University Press.

Chattopadhyaya, D. (Ed.) (1990). *Carvaka/Lokayata*. New Delhi: Indian Council of Philosophical Research.

Dasgupta, S. N. (1930). *Yoga Philosophy*. Delhi: Motilal Banarsidass.

Eliade, M. (1954). *The Myth of the Eternal Return*. Princeton, NJ: Princeton University Press.

Eliade, M. (1969). *Yoga: Immortality and Freedom*. Princeton, NJ: Princeton University Press.

Kakar, S. (1981). *The Inner World: A Psychoanalytic Study of Childhood and Society in India.* Delhi: Oxford University Press.

Kakar, S. (1991). *The Analyst and the Mystic.* Chicago, IL: University of Chicago Press.

Kane, P. V. (1974). *History of the Dharmasastra: Ancient and Medieval Religious and Civil Law.* Poona, India: Bhandarkar Oriental Institute.

Larson, G. J., & Bhattacharya, R. S. (Eds.) (1987). *Encyclopedia of Indian Philosophies—Samkhya.* New Delhi: Motilal Banarsidass.

Loewald, H. (1972). The experience of time. *Psychoanalytic Study of the Child, 27*: 401–410.

Michaels, A. (2004). *Hinduism: Past and Present.* Princeton, NJ: Princeton University Press.

Parsons, W. (1999). *The Enigma of the Oceanic Feeling.* Oxford: Oxford University Press.

Peters, F. E. (2003). *The Monotheists.* Princeton, NJ: Princeton University Press.

Radhakrishnan, S. (1939). *Eastern Religions and Western Thought.* Oxford: Oxford University Press.

Radhakrishnan, S. (Ed.) (1953). *The Principal Upanishads.* Delhi: Oxford University Press.

Reddy, S. (2005). Psychoanalytic process in the sacred Hindu text, *The Bhagavad Gita.* In: S. Akhtar (Ed.), *Freud on the Ganges* (pp. 309–333). New York: Other Press.

Rolland, R. (1984). *The Life of Ramahrishna.* Calcutta: Advaita Ashrama.

van Buitenen, J. A. B. (1981). *The* Bhagavad Gita *in the Mahabharata.* Chicago, IL: University of Chicago Press.

Waelder, R. (1963). Psychic determinism and the possibility of predictions. *Psychoanalytic Quarterly, 32*:14–42.

Zimmer, H. (1974). *Myths and Symbols in Indian Art and Civilization.* Princeton, NJ: Princeton University Press.

Zimmer, H. (1989). *The Philosophies of India.* Princeton, NJ: Princeton University Press.

INDEX